Land Snails

of British Columbia

Robert G. Forsyth

ROYAL **BC** MUSEUM

Victoria, Canada

Published by the Royal BC Museum,
675 Belleville Street, Victoria, British Columbia, Canada, V8W 9W2
Web site: wwwroyalbcmuseum.bc.ca

Printed in Canada.

Front-cover photograph of *Cryptomastix germana* and back-cover
photograph of *Ariolimax columbianus* by Kristiina Ovaska.

National Library of Canada Cataloguing in Publication Data
Forsyth, Robert G.
    Land snails of British Columbia

    (Royal BC Museum handbook)

    At head of title: Royal BC Museum handbook.
    Includes bibliographical references and index.
    ISBN 0-7726-5218-X

    1. Snails - British Columbia. 2. Slugs (Mollusks) -
British Columbia. 3. Gastropoda - Classification.
I. Royal BC Museum. II. Title.

QL430.4F67 2004   594.3   C2004-960094-X

CONTENTS

Preface   1
Introduction   3
  Where Snails and Slugs Live   6
  Reproduction and Life History   7
  Diet, Movement and Defence   10
  The Shell   13
  The Fauna in British Columbia   17
  Exotic Snails and Slugs   19
  Names of Snails and Slugs   20
  History of Terrestrial Malacology in B.C.   21
  Important Literature   22
Checklist   23
Identification and Keys to Genera   29
  Identifying Snails and Slugs   29
  Key to Snails   30
  Key to Slugs   33
Species Accounts   35
  Family Carychiidae   36
  Family Succineidae   39
  Family Cionellidae   41
  Family Lauriidae   42
  Family Valloniidae   44
  Family Pupillidae   51
  Family Vertiginidae   53
  Family Haplotrematidae   70
  Family Testacellidae   74
  Family Punctidae   75

Family Discidae   79
Family Oreohelicidae   84
Family Pristilomatidae   87
Family Euconulidae   94
Family Gastrodontidae   97
Family Daudebardiidae   101
Family Milacidae   108
Family Vitrinidae   109
Family Boettgerillidae   110
Family Limacidae   112
Family Agriolimacidae   116
Family Arionidae   123
Family Polygyridae   148
Family Thysanophoridae   156
Family Bradybaenidae   159
Family Helicidae   160
References and Bibliography   164
Glossary   179
    Scientific Names of Plants and Other Animals   182
Acknowledgements   183
Species Index   185

PREFACE

This handbook describes and illustrates the 92 species of land snails and slugs known to live in British Columbia. It is the first comprehensive publication on the snail and slug fauna of the province (aside from a few early checklists and short papers), and its arrival is timely, given the increasing interest among biologists in the biodiversity of the province and a general public demand for more information on their natural surroundings. Furthermore, many taxonomic changes have occurred since the last major work on North American land snails and slugs was published over 50 years ago.

The technical names for these animals are *molluscs* or *gastropods*, and I use these terms throughout the text. I wrote this book primarily for non-specialists, so I simplified the terminology as much as possible and included a glossary of the more specialized words. For biologists and others who may wish to delve deeper into the malacological literature, I provide a list of selected references after each section and species account, and an extensive bibliography at the back of the book.

My interest in molluscs formed at an early age and grew as I learned more about them over the years. As my studies intensified, I discovered an astonishing lack of information on the terrestrial snails and slugs of B.C., and so decided to write this handbook. Like all Royal BC Museum handbooks, it is an amalgamation of information from published literature, my own observations, scientific collections and personal communications with biologists and naturalists (see the Acknowledgements on page 183). In many cases the distributions in the species accounts are based on unpublished locality records in the invertebrate collection of the Royal BC Museum or in my own study collection.

This book is a work in progress. The taxonomy is not fully resolved for several of our species (pointed out in the text), and because B.C. is so large and so little of it has been explored for land snails and slugs, the searching of new localities and unusual habitats almost always yields new information.

There is still much to be learned about the land snails and slugs of British Columbia, and I hope that this handbook will encourage further study.

Smithers, British Columbia
March 2004

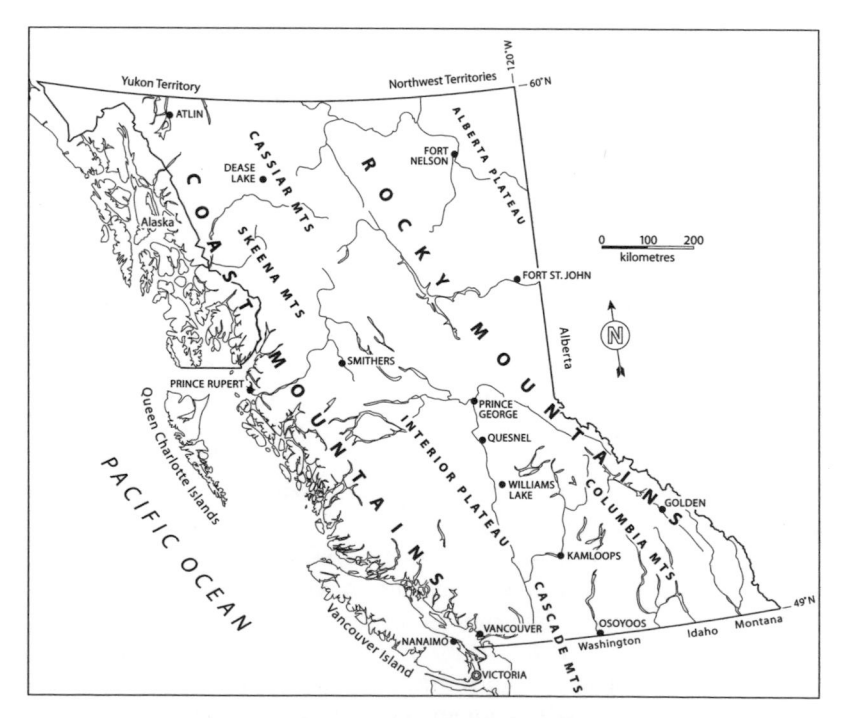

Figure 1. General geographic features of British Columbia.

INTRODUCTION

Land snails and slugs have a bad reputation. Most people dismiss them as slimy, slow-moving creatures seldom worth a second thought. Because they are small and reclusive, we seldom notice them, unless they become pests in our gardens. But from an evolutionary point of view, slugs and snails are highly successful. And they are important components of the ecosystems in which they live. There may be as many as 35,000 species of land snails and slugs worldwide, more than all the mammals, birds, reptiles and amphibians combined. Snails and slugs are well adapted to live virtually everywhere and occupy a variety of environments, including high arctic and alpine tundra, grasslands, temperate and tropical forests, and even the driest deserts.

Land snails and slugs are a terrestrial offshoot of a mostly marine group of unsegmented, soft-bodied organisms, known as *molluscs* – (from the Greek *mollis,* meaning "soft"). Scientists classify this diverse group of animals as the phylum Mollusca. Within the phylum, they recognize seven major subdivisions, the classes Polyplacophora, Aplacophora, Gastropoda, Bivalvia, Scaphopoda, Monoplacophora and Cephalopoda. All land snails and slugs belong to the largest class, Gastropoda. The word *gastropod* is a compound of two Greek words meaning "stomach-foot", referring to the position of the visceral mass on top of the foot, a muscular organ of locomotion.

Gastropods have a well-defined head that bears one or two pairs of tentacles (figures 2, 3). Sensory tentacles have chemoreceptors and the longer ocular tentacles (when present) have eyespots, capable of detecting light, shadow and movement. On the head,

Figure 2. A typical snail
(*Haplotrema vancouverense*)
showing head, tentacles and foot.

usually near the right ocular tentacle is a small opening to the reproductive tract. A fold of tissue, called the mantle, envelops and protects the visceral mass, secretes a shell, and forms a lung for respiration (or, in mostly aquatic snails, has gills).

Slugs lack an obvious shell, but their mantle is conspicuous as a raised area behind the head (figure 3). In snails, the mantle is less prominent and since it lines the inner surface of the shell, it usually protrudes from the shell's opening as a narrow fold.

The mouth is on the lower surface of the head and opens to the buccal cavity, which bears the radula, an organ unique to molluscs. In its typical form, the radula is a horny, ribbon-like structure possessing multiple rows of tiny teeth.

All the land snails and slugs described in this book belong to the Order Pulmonata (in some classifications, Pulmonata is ranked as a class, but many recent classification schemes list it as an order). Pulmonates are an evolutionarily successful group of hermaphroditic molluscs characterized by a vascular sac in the mantle cavity, which functions as a lung for breathing air; a rounded hole on the right side of the body (figure 4), called the pneumostome, leads into the mantle cavity. But not all terrestrial gastropods in the world are pulmonates. Elsewhere, especially in the tropics, there are terrestrial prosobranchs, gastropods furnished with gills for respiration.

The evolutionary history and classification of the Gastropoda and Pulmonata are controversial and not fully resolved. Recent

Figure 3.
A typical slug
(*Arion subfuscus*)
showing the mantle, pneumostome and tentacles.

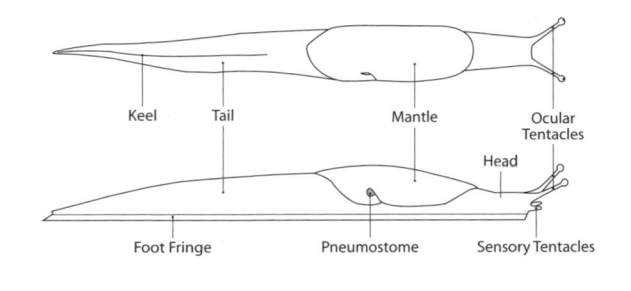

Figure 4. General features of a slug.

Keel   Tail   Mantle   Ocular Tentacles

Head

Foot Fringe   Pneumostome   Sensory Tentacles

advances have brought about the modification, dismantling, abandonment and creation of some high-ranking groups within the Gastropoda – and different opinions among taxonomists. Because this book is foremost a field guide to the fauna of British Columbia and not a treatise on molluscan systematics, I will not dwell on the taxonomic issues of the class.

The species in this book represent two major subdivisions of the Pulmonata. The suborder Stylommatophora – by far the most successful group of Pulmonata – includes 90 of the 92 species described in this book. Species in this suborder share two major characteristics: the male and female reproductive tracts open into a common chamber and exit through a single genital pore; and there are usually two pairs of tentacles with eyespots at the tips of the posterior (upper) pair. The tentacles are *invaginal* – they retract in on themselves, like the fingers of a rubber glove when you pull out your hand. Stylommatophorans may have an obvious coiled shell (the snails), or they may have a tiny shell or no shell at all (the slugs). The other two species in this book belong to the suborder Acteophila, characterized by these traits: the male and female reproductive systems open externally by way of two separate genital pores, and there is one pair of non-invaginal tentacles with eyespots at their bases. No species of Acteophila are slug-like in form.

Selected references: Dutra-Clarke et al. 2001, Emberton et al. 1990, Ponder and Lindberg 1996, Roth and Sadeghian 2003, Solem 1984, Thollesson 1999, Wade and Mordan 2000, Wade et al. 2001, Yoon and Kim 2000.

Where Snails and Slugs Live

Land snails and slugs live in almost all terrestrial environments, and few places are entirely without these animals. In British Columbia, they live in all types of forests, grasslands, marshes and alpine tundra, on talus slopes, and on open ground. Many species, particularly exotic ones (see page 19), are common in gardens, pastures, roadsides and other disturbed sites.

The overall distribution, diversity and abundance of native snails and slugs in the province are affected by a combination of climatic, geological and biological factors. All snails and slugs require adequate moisture and shelter, as well as a supply of food. In the western U.S. and Canada, scientists have demonstrated a strong link between the type and abundance of plant cover and snail diversity. The depth of leaf litter, the amount of organic material in the soil and the soil's ability to retain moisture are important factors; land snails are usually more abundant and diverse in places with deep leaf litter. The amount of calcium, a necessity for building shells, also affects local snail abundance and diversity.

The physical form of the land – the angle and aspect of the slope and its elevation – also influence land-snail diversity and abundance, but physiographic features and the type and abundance of vegetation are closely tied to the amount of annual precipitation. In 2001, I compared valley and mountain faunas in the Hazelton and Skeena ranges of northwestern British Columbia, and found that the diversity of land snail species was greater at wetter, more elevated sites than at drier valley sites. This corresponded to greater precipitation on the higher mountainsides, and proved to be an exception to the general rule that diversity decreases with elevation.

In British Columbia the humid, deciduous or mixed-wood forests of the coast and the wet interior mountains have the highest land-snail diversity. On the other hand, purely coniferous forests look as if they have much less species diversity, but as speculated by E.J. Karlin (1961), low-density populations are likely present for most species in these forests.

Although land snails and slugs are remarkable for occupying a wide range of habitats, they are susceptible to desiccation and most seek out the shelter of moist places found under boards, rocks, dead wood, plants and fallen leaves. Many are nocturnal and venture out in daylight only during wet weather. Very small snails spend their entire life in leaf litter on the forest floor, moving up and down in the litter layer depending on moisture conditions. To avoid summer drought, terrestrial gastropods may aestivate to reduce the rate of water loss, and to cope with winter cold, many hibernate in sheltered places. During aestivation and hibernation, these animals lower their heart rate to a minimum and greatly reduce their oxygen consumption; many will seal their shell opening with a mucus sheet called an epiphragm.

Most species become more active (and more likely to be seen) at certain times of the year, usually in late spring or late summer and fall when the weather is wet and not overly cool.

Selected references: Boag 1985, Boag and Wishart 1982, Cameron 1986, Forsyth 2001d, Hoff 1962, Karlin 1961, Kralka 1986, Myers 1972, Ports 1996.

Reproduction and Life History

Normally, pulmonate gastropods are simultaneous hermaphrodites – each individual produces both sperm and eggs. In most species, under typical conditions, individuals must copulate and exchange sperm, but some species have evolved to be partially or wholly self-fertilizing. Individuals of most land snails and slugs have both male and female genitalia (figure 5), but in some, the male organs are reduced or absent, making them functionally females.

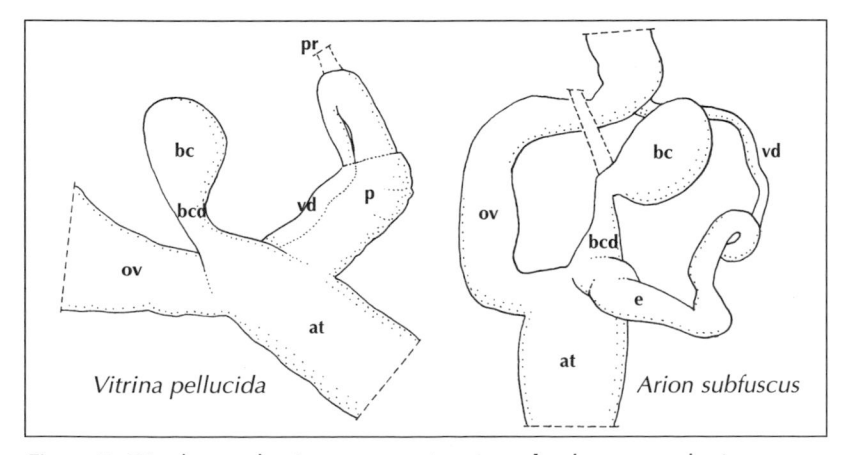

Figure 5. Distal reproductive system: **at**, atrium; **bc**, bursa copulatrix; **bcd**, bursa copulatrix duct; **e**, epiphallus; **p**, penis; **ov**, oviduct; **pr**, penis retractor; **vd**, vas deferens.

The morphological characters of the pulmonate reproductive system are often key in determining species, genera and higher taxa, so it is useful to understand its basic anatomy. The normal condition is for the reproductive system to have both male and female organs. Eggs and sperm are formed at the far end in the structure known as the ovotestis. From there, they travel down the hermaphroditic duct to the fertilization-pouch/bursa-copulatrix complex (carrefour complex), a small structure in relation to the rest of the system, imbedded in the albumen gland. In primitive land snails (Acteophila) the sperm duct and oviduct are fully separate. In stylommatophorans these ducts are grooves in a single structure, the spermoviduct (sometimes called the common duct); only at the distal end of the spermoviduct do the male and female ducts completely separate.

Up to this point, the reproductive system is relatively undifferentiated among species, but what follows – sometimes called the distal reproductive system (figure 5) – is more complex and more specific to species. The male part of the distal reproductive system includes the vas deferens, which leads into a dilated portion called the epiphallus or directly into the penis. The penis (absent in *Arion* species) is a muscular organ that is eversible (capable of turning inside out) and may have accessory structures (folds, a stimulator, flagellae), important for mate recognition and stimulation. The main female part of the distal system is the oviduct, which may also have associated structures such as a dart sac and branched mucus glands. In stylommatophorans the oviduct and penis (or epiphallus, if the penis is absent) connect to a common chamber called the atrium before the reproductive system exits the body wall though the genital pore; in more primitive snails, male and female organs have separate openings. Another major structure of the distal genitalia is the bursa copulatrix (sometimes called the spermatheca), a sac-like structure that receives and reabsorbs excess spermatazoa and other reproductive products; the bursa copulatrix duct opens into either to the atrium or the oviduct.

In *Zonitoides nitidus*, some species of *Vertigo* and *Arion*, and some populations of *Deroceras laeve*, self-fertilization is the primary means of reproduction and cross-fertilization relatively rare. Parthenogenesis – the development of an unfertilized egg – does not occur in terrestrial gastropods, except for possibly *D. laeve* (and there is good evidence against this). Often associated with self-fertilization is the reduction or loss of the male genital organs in a portion of a popu-

Figure 6. Courtship behaviour of two *Prophysaon andersoni* prior to coitus.

lation. Three degrees of development may be present: euphallic individuals have the "normal" state where both male and female organs are fully developed; hemiphallic individuals have the male organs reduced in size; and aphallic animals lack the male portion of the reproductive system altogether. In some terrestrial gastropods – *D. laeve* or some *Vertigo* species for example – aphally is more prevalent than euphally.

Many self-fertilizing snails are very small. Finding a mate may be difficult or impossible if the population density is low and distances between individuals are great or unfavourable for travel. Self-fertilization allows a population to boost its density and colonize new territory in the absence of a mate. Some normally cross-fertilizing species may resort to self-fertilization if a mate has been unavailable for some time, but the resulting eggs are fewer in number and less viable.

Many species of terrestrial snails and slugs engage in elaborate courtship behaviours to help identify suitable mates: partners will encircle each other, touching and biting or licking (figure 6); they secrete sexual pheromones in mucus; and many have matching structures in the distal genitalia, acting like a lock and key, which isolates them reproductively. In a strange play of acrobatics, pairs of *Limax maximus* mate at night while suspended mid-air from overhanging surfaces on a thread of mucus.

Most snails and slugs are oviparous and lay eggs singly or in clusters in depressions in the ground or amid leaf litter. Eggs are roughly spherical to oval, white to yellowish, and soft or hardened

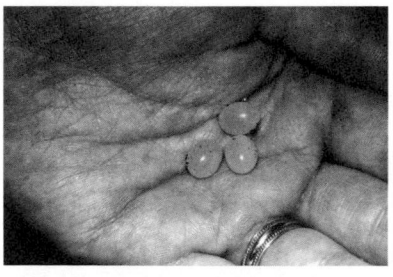

Figure 7. Eggs in the hand: the large, oval eggs of *Ariolimax columbianus*.

with calcium (figure 7). *Zoogenetes harpa* and species of *Oreohelix* are examples of ovoviviparous snails – parents retain their eggs inside, protecting them until they hatch.

We do not know how long most snails live. In general, tiny species are thought to be short-lived – hatching, maturing, reproducing and dying within a year. Other factors limiting the life spans of land gastropods are: the lack of an external shell, having a semitransparent shell, a sunny habitat and a stable environment that favours annual reproduction. There are some land snails that live relatively long lives: a species of *Monadenia* lives for at least 8 years, and the Middle Eastern land snail, *Cristataria genezarethana*, lives at least 16 years, reaching maturity after about 11.

Selected references: Barker 2001; Bayne 1973; Gómez 2002; Heller 1990, 1993; Heller and Dolev 1994; Hyman 1967; Lebovitz 1998; McCracken and Selander 1980; Morton 1979; Nicklas and Hoffman 1981; Pokryszko 1987b; Solem 1974; South 1992; Tompa 1979, 1984; Walton 1970.

Diet, Movement and Defence

Land snails and slugs eat fungi, organic detritus and plants, living or dead. Many feed on carrion as well and some are predacious. We know little about the specific diets of most terrestrial gastropods in British Columbia, but we think most of our species eat fungi and detritus.

Snails and slugs propel themselves forward with a series of wave-like muscular contractions along the length of the foot. Compared to other terrestrial animals, gastropods do not move efficiently; they require considerably more energy, mostly in the production of mucus. And because mucus production requires a lot of water, snails and slugs must limit their movement during dry weather; most of them aestivate. The slug *Ariolimax columbianus* secretes a strong, elastic mucus from the caudal mucus pore at the tip of the tail, allowing it to descend from vegetation on a mucus thread.

Several species have a homing ability, enabling them to return to a favourable shelter after the nightly raid on the local vegetable patch. Homing has been attributed to simply following a trail, but *Limax maximus* – and likely other species – can find their way home by chemical stimuli (smell).

Terrestrial gastropods travel literally at a snail's pace, which places severe limits on an individual's range. A. Solem (1974) estimated that the movements of a land snail during its lifetime are confined to a six-metre radius, although it is likely that some species cover greater distances and others much less depending on environmental conditions and habits. Passive dispersal is an important factor in species' distribution – floodwaters, wind, mammals (including humans), birds, and even insects may transport terrestrial gastropods or their eggs over great distances, thereby establishing new populations; and gravity may cause snails to roll down mountain slopes.

Many animals prey on snails and slugs, including birds, small mammals, amphibians, reptiles, other terrestrial gastropods, carabid ground beetles and many other insects. To cope with predators, snails and slugs have evolved various defence strategies. Land snails gain at least some protection from predation by withdrawing into their shells. Apertural barriers (see figure 15) further narrow the shell's opening and may deter a predator. Certain species also exude disagreeable odours when irritated: *Oxychilus alliarius* produces mucus that smells strongly of garlic, which is believed to deter predators .

In contrast to snails, slugs gain little or no protection from their shells. Perhaps because of this, slugs seem to show some of the most highly adapted anti-predation responses. Mucus can be discharged in large quantities to foul the mouthparts of predatory beetles. Some slugs, such as *Ariolimax columbianus*, exude a dried plug of mucus from a specialized gland at the tip of the tail. Scientists believe that a predator bites off the plug and is temporarily occupied while the slug escapes.

Two groups of British Columbia slugs have rather unusual defences. Jumping slugs (*Hemphillia* species) are so named for their violent writhing, twisting and swinging of the tail when provoked. *Hemphillia* curls the tip of its slender tail forward to touch its head when at rest, and from this position, it can quickly recoil the tail and

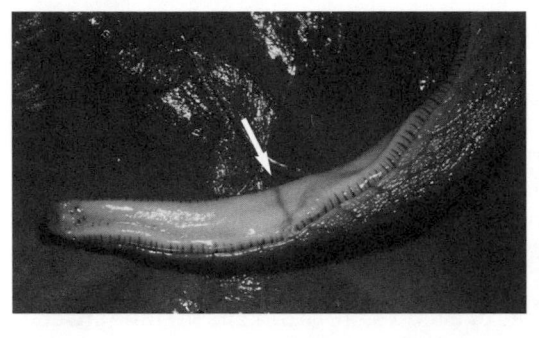

Figure 8. The site of autotomy in *Prophysaon foliolatum* is marked by a dark line on the sole of the foot.

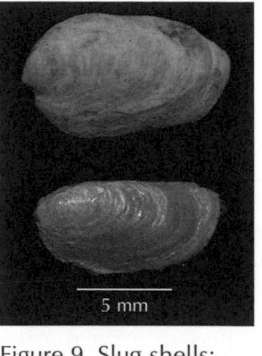

Figure 9. Slug shells: *Ariolimax columbianus* (above) and *Limax maximus*.

flip itself a short distance. The tail droppers (*Prophysaon* species) are capable of autotomy (self-amputation by spontaneously casting off the tail) when attacked by a predator. The place on the tail where autotomy takes place is marked by an oblique constriction, sometimes evident as a dark line on the sole of the foot (figure 8). *Prophysaon foliolatum* casts off its tail in two to five seconds after it feels a predator's attack, such as the bite of a beetle: the autotomy zone contracts, the tail section swells slightly, the body in front of the autotomy zone secrets large amounts of sticky mucus, and then the tail is cast off. Only an attack at or behind the autotomy zone stimulates the slug to cast its tail, suggesting that the stickier mucus discourages predators and redirects them to the tail section where the mucus is not as sticky. Experiments showed that an individual could discard and regenerate its tail more than once. In the laboratory, *Prophysaon* regenerated a new tail within five weeks. Other slugs, including *Deroceras reticulatum*, *Limax maximus* and species of *Lehmannia*, also self-amputate their tails, but we know less about the process for them.

Selected references: Adam 1960, Anderson and Mehl 1992, Baur et al. 1997, Branson 1977, Denny 1980, Deyrup-Olsen et al. 1986, Dundee et al. 1967, Elves 1961, Gelperin 1974, Hand and Ingram 1950, Kirchner et al. 1997, Lissmann 1945, Lloyd 1970a, Pakarinen 1994a, Peake 1978, Pilsbry 1948, Rees 1965, Richter 1980, Solem 1972, Stasek 1967, Trueman 1983.

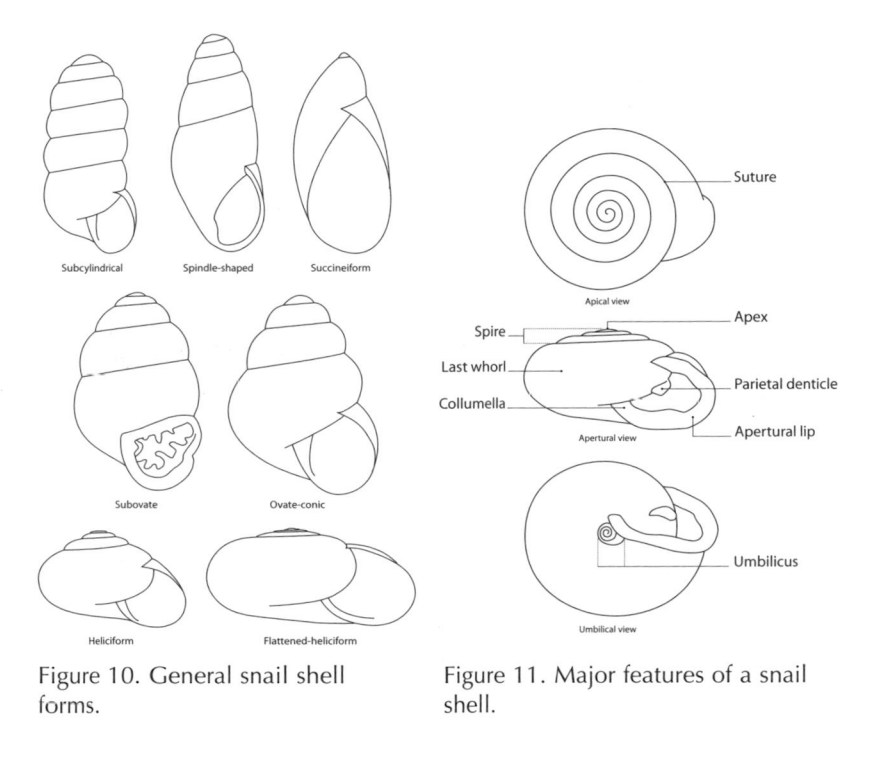

Figure 10. General snail shell forms.

Figure 11. Major features of a snail shell.

The Shell

A snail's shell is a non-living structure (like a fingernail) usually shaped like a tube coiled around a central axis. Most slugs have a flat internal shell (figure 9) – or in *Arion* species, a few isolated shell granules – offering little or no protection. But the shell is a major feature of most land snails, and most snails can completely retract into it for protection against predators and water loss. Snail shells are diverse in form. For convenience, I have classified the shell forms of all the species in this book into seven types (figure 10).

The shell of a snail (figure 11) consists of an outer organic layer called the periostracum and an inner, largely calcium carbonate layer. These layers are formed by incremental additions laid down by the edge of the mantle. Each full coil of the tube is called a whorl. The first part of the shell that forms, normally before the snail hatches from the egg, is the protoconch, which is often marked with unique surface ornamentation or sculpture (see SEM images on page 98). All subsequent whorls of the shell make up the teleoconch.

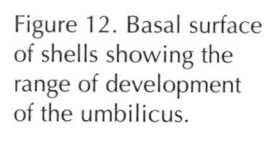

Figure 12. Basal surface of shells showing the range of development of the umbilicus.

Umbilicus absent   Umbilicus pit-like   Umbilicus prominent

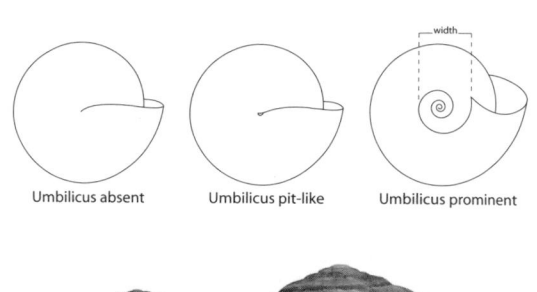

Figure 13.
Growth series of
Monadenia fidelis.

The spire is formed from all whorls, except for the last; the tip of the spire is the apex. Whorls increase in size as the snail grows to adulthood; where each whorl is attached to the next, a spiral line, called the suture, is formed. The opening in the last whorl, through which the animal extends its body, is the aperture. The central pillar of the shell, called the columella, may be solid or hollow. If it is hollow, there is an umbilicus, a hole or depression that the whorls coil around (figure 12).

Snails have a limited ability to repair and thicken their shell from within. All shell expansion occurs at the apertural lip. The shells of most land snails exhibit allometric growth; that is, they change shape as they increase in size and reach maturity (figure 13). Many snails have a final growth phase when they reach sexual maturity. This is marked by a change in the direction or expansion rate of coiling, or the development of a thickened, flared, indented or recurved apertural lip (figure 14). Often along with the development of the apertural lip comes the formation of apertural denticles (figure 15).

With a few exceptions, apertural denticles are developed only in adults. Their purpose remains unclear, and they have evolved independently in many groups. Apertural denticles may constrict the aperture and offer protection against predators, but they may also help decrease water loss

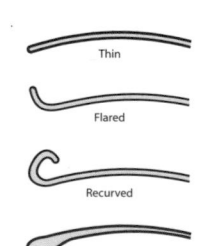

Figure 14. Variations in the development of the apertural lip of snail shells.

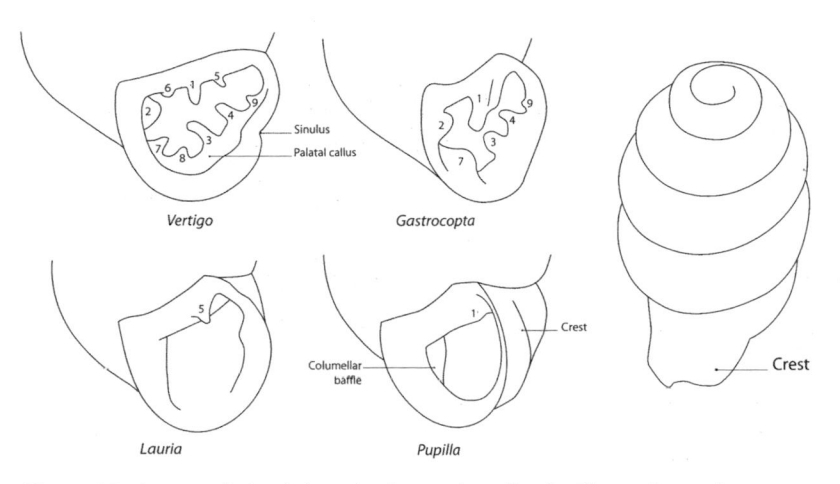

Figure 15. Apertural denticles, sinulus, columellar baffle and crest in genera of the families Vertiginidae and Pupillidae. The denticles are: 1, parietal; 2, columellar; 3, lower palatal; 4, upper palatal; 5, angular; 6, infraparietal; 7, subcolumellar; 8, infrapalatal; 9, suprapalatal. The palatal denticles may rest on a thickened palatal callus. In *Gastrocopta*, the parietal is fused with the angular; in *Pupilla*, the parietal may be very low or absent.

through evaporation, support the snail's internal organs, position the shell during movement, prevent shell breakage by strengthening its edge or even prevent drowning by trapping air inside the shell and making it more buoyant. The number, form and position of apertural denticles can be important in species identification.

A few groups in this book (especially the families Vertiginidae and Pupillidae) possess denticles and some unique features on the apertural lip (figure 15). The four main denticles present in most of our *Vertigo* species and some related genera are named for their position inside the aperture: columellar, on the columella; parietal, on the wall of the last whorl; and upper and lower palatals, inside the apertural lip. Some species have more denticles: angular, above the parietal denticles in the angle formed by the apertural lip; infraparietal, below the parietal; suprapalatal, above the upper palatal; infrapalatal, below the lower palatals; and subcolumellar, below the columellar, almost basal. In other literature, authors have called these other denticles "plicae" (or "folds") and "lamellae" (or "plates"), but for simplicity, I am using "denticle" to describe any of the projections inside the aperture. *Pupilla* species have a columellar

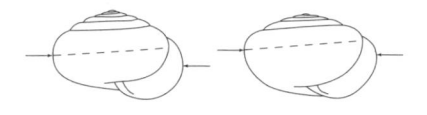

Figure 16. Relative position of the periphery on the last whorl. *Left*, medial on the whorl; *right*, above the midline of the whorl.

baffle behind the columella. Some *Vertigo* species have a sinulus, an embayment in the curvature of the apertural lip (called the auricle, especially when greatly indented). Some Pupillidae, Vertiginidae and Polygyridae species have a crest (figure 15) – a raised, axial bulge on the last whorl behind and parallel to the apertural lip, usually set off by a constriction.

Viewing the shell from the side (perpendicular to the shell axis), the whorls may be flat or convex in profile. The part of a whorl farthest from the axis of a spiral shell that whorl's periphery. The periphery may be rounded or angular, and is at, above or (less frequently) below the middle of the whorl (figure 16).

Snail shells may be opaque or translucent, and glossy or dull. They are uniformly coloured or display a pattern, often spiral bands. Most shells are also ornamented with raised or incised sculpture (figure 17). Large, raised, rounded ridges are ribs, and smaller ones riblets; thin, blade-like ridges are lamellar ribs. Finally, for an irregularly indented surface, like hammered metal, I use the term "dimpled sculpture". Both sculpture and pigmentation patterns can be described as axial (running approximately parallel to the central axis) or spiral (running parallel to the direction of coiling).

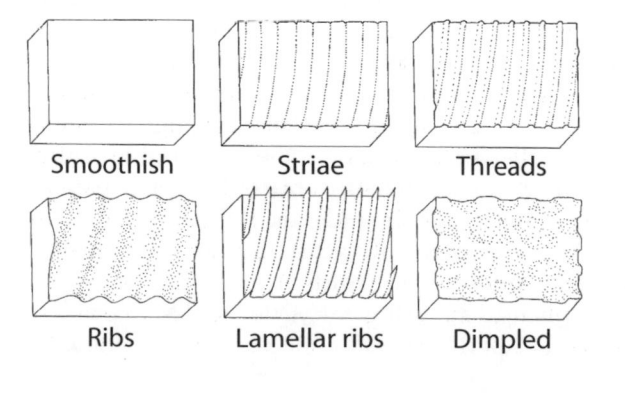

Figure 17. Surface sculpture of land snails.

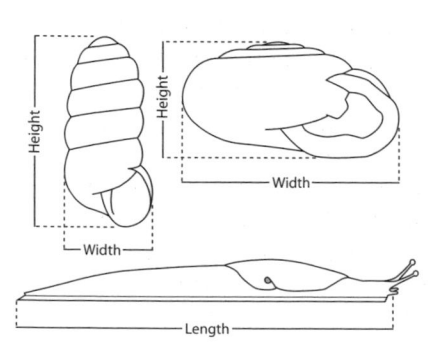

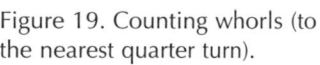

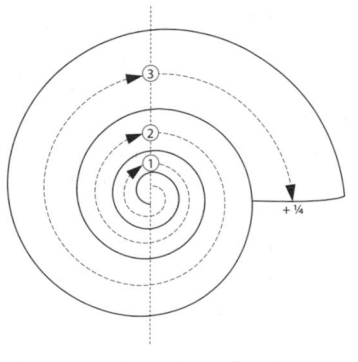

Figure 18. Measuring land snails and slugs.

Figure 19. Counting whorls (to the nearest quarter turn).

For this book, I have measured shells by either their width or their height, whichever is larger (figure 18). The width (sometimes breadth or diameter) is the maximum measurement of the shell at right angles to the central axis. The shell height (or length) is the maximum measurement along the central axis. These two terms merely refer to the shell and not its orientation on the living animal. The body length of slugs is measured when the animal is fully extended and active.

It is often useful to describe shells by the number of whorls that they have. Here I have counted the number of whorls to the nearest quarter turn (figure 19).

Selected references: Emberton 1995; Kerney and Cameron (1979); Pokryszko 1990, 1997; Solem 1972, 1974; Watabe 1988.

The Fauna in British Columbia

Temperate North America is divided into three major faunal groups of terrestrial gastropods: boreal, western and eastern. The eastern and western faunal divisions are more-or-less separated by the Rocky Mountains, both merging gradually into a boreal faunal division in the north. Some malacologists have subdivided the western division into faunal provinces, but these are difficult to recognize in British Columbia. Most species in this book have western or boreal affiliations; some species typical of the eastern North American fauna, such as *Gastrocopta holzingeri*, are rare in B.C.

British Columbia is over 950,000 square kilometres in area and is topographically diverse. Patterns of precipitation, geologic history and topography have created biologically distinct regions in the province. The distribution of snails and slugs follows these patterns, but many species are much more widespread.

The south coast has more species diversity than any other part of B.C.: *Allogona townsendiana, Monadenia fidelis, Ariolimax columbianus, Cryptomastix germana, Prophysaon foliolatum, Vespericola columbianus* and several minute litter-dwelling species. The distributions of these species follow the coast northward, with some extending up the coast of Alaska.

These coastal species penetrate inland up major river valleys, and a few species (such as *Striatura pugetensis*) also live in wet forests of the Cariboos and Rockies. The area between the coast and the wet interior belt is the relatively dry Interior Plateau. This vast region and much of northern B.C. have an impoverished fauna characterized by the likes of *Euconulus fulvus, Discus whitneyi, Zonitoides arboreus* and other species that are generally distributed throughout the province.

The southeastern portion of B.C. is typified by species extending northward in the Rocky Mountains and along the Columbia River drainage into Canada from the United States. In the mountains, we find species of *Oreohelix, Magnipelta mycophaga* and *Microphysula ingersollii*, while in the valleys are *Allogona ptychophora, Anguispira kochi* and *Cryptomastix mullani* in addition to *Oreohelix*.

In the northeast, the Peace River region – east of the Rocky Mountains and part of the Alberta Plateau – shares the same fauna as that of the Interior Plateau and the northern mountains.

Currently, there are no named species of land snails or slugs with a range exclusive to B.C.

Conservation of land snails and slugs in Canada has only just begun. *Cryptomastix devia*, a species that was collected in B.C. only a few times in many years, is now believed to be extirpated; other species are threatened. The most serious threat today to the survival of terrestrial gastropod species in B.C. is habitat loss caused by human activities; globally, predation by introduced species is also a major concern.

Selected references: Henderson 1931, Peake 1978, Pilsbry 1948.

Exotic Snails and Slugs

Many snails and slugs prosper in habitats modified by humans. Through passive dispersal in agricultural products and shipping containers, and by other means, these species have expanded their ranges to all continents except Antarctica. At least 25 exotic species of terrestrial gastropods have colonized British Columbia, more than a third of the species in this book. Exotic snails and slugs likely arrived in British Columbia with the first European settlers, and by the late 1880s the Grey Fieldslug, *Deroceras reticulatum*, was already in Victoria gardens.

Most exotic snails and slugs in British Columbia originated in Europe, but their precise origin, date of arrival and means of introduction are unknown. In recent years, some larger snails and slugs have been intercepted on imported trees and shrubs, suggesting that other smaller species and eggs are also transported this way. Many garden centres and nurseries are richly populated by exotic gastropods. Snails, slugs and their eggs are transported to new areas with soil, garden debris, rocks, wood and other materials.

A few of these snails and slugs are familiar pests (figure 20), but the majority go unnoticed by most people. All are synanthropic – closely associated with humans and man-made landscapes. These opportunistic creatures thrive in unnatural and disturbed habitats. Some exotic species in this book are widespread and common, while others have more limited ranges in B.C., possibly because the arrived more recently or they are not well suited to our climate.

Few native species of snails and slugs are serious agricultural pests, although native slugs often frequent gardens adjoining natural areas (*Prophysaon andersonii*, for example, is sometimes regarded as a nuisance). In British Columbia, the terrestrial gastropods that do the most damage to plants are exotic species; *Deroceras reticulatum* and the several larger species of *Arion* are probably the most important pests.

Also of concern is the potential long-term impact of exotic species that could prey on the

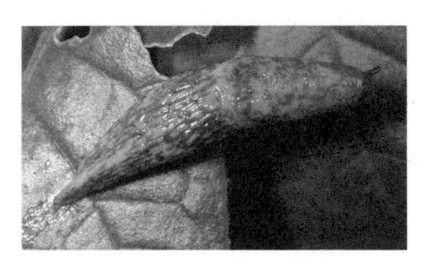

Figure 20. The Grey Fieldslug, *Deroceras reticulatum*, is an important pest species, shown here on a potato plant.

native fauna. Some species, such as *Oxychilus draparnaudi* and *Aegopinella nitidula*, could pose a problem should they expand their ranges into natural areas.

Selected references: Barker 2002, Forsyth 1999, Forsyth et al. 2001, Frest and Rhodes 1982, Godan 1983, Hanna 1966, Robinson 1999, Roth and Pearce 1984, South 1992, Taylor 1889.

Names of Snails and Slugs

The scientific naming of animals revolves around the recognition of species. Most biologists define a species as a group of individuals capable of breeding with each other but not with individuals from other such groups. In practice, scientists usually distinguish species by morphological differences (as used in the descriptions in this book).

A biologist who finds and describes a species gives it a unique scientific name consisting of two words. The first word, the genus name, is always written with the first letter capitalized; the second word, designating the species, is never capitalized. Scientific names of species are always binomial – that is, both words must be used. They are written in italics, because the names are based on Latin or Greek, and as in the Latin language, the gender of the species or subspecies names must match the gender of the genus.

Often added to the scientific name – especially in technical publications – is the name of the person who first described the species and the date it was published. For example, Henry Pilsbry described *Vertigo andrusiana* in 1899, so we write "*Vertigo andrusiana* Pilsbry, 1899". In cases where the species has been subsequently moved from the genus in which it was originally described, we enclose the author's name and the date in parentheses, as in *Cryptomatix germana* (Gould, 1851). Augustus A. Gould described this species as *Helix germana* in 1851, and later the species was moved to the genus *Cryptomatix*. To be perfectly accurate, because Gould described the species in a publication by Amos Binney, we can write the name in its long form, "*Cryptomastix germana* (Gould *in* Binney, 1851)".

In large and diverse genera, relationships are shown by grouping related species into subgenera. The typical subgenus takes the same name as its genus. (*Cryptomastix* and *Micranepsia* are two subgenera of *Cryptomastix*.) Although not a requirement, subgenera can be written as part of the full name of an animal: "*Cryptomastix (Micranepsia) germana*"; "*C. (Cryptomastix) mullani* (Bland & Cooper, 1861)". Note that, for convenience, we can abbreviate the genus name once it has been established.

Sometimes, below the species, is another less-inclusive unit called the subspecies. Subspecies are populations within a species that differ in morphology and are geographically or ecologically separated from each other. *Cryptomastix mullani olneyae* is a subspecies of *C. mullani*; the typical subspecies is *C. mullani mullani.*

Many species also have common names, given to them by people using their native language, often different from one region to the next. For example, *Euconulus fulvus* is called Tawny Snail or Brown Hive in English, Tolslakje in Dutch, and Helles Kegelchen in German. Scientific names, on the other hand, are unique and can be used worldwide by everyone no matter what language you speak.

History of Terrestrial Malacology in British Columbia

The history of terrestrial malacology in British Columbia is short and features relatively few participants. The first scientific collections of terrestrial molluscs in British Columbia were made by John Keast Lord, naturalist for the British North America Boundary Commission (1858–62); his specimens are now in the British Museum, London. On various expeditions by the Geological Survey of Canada, George M. Dawson, J. Macoun, John Spreadborough and J.B. Tyrrel made early collections of B.C. land snails, which are now housed in the Canadian Museum of Nature.

Others have travelled and collected in British Columbia since the late 1800s. P. Brooks Randolph – for a time, the zookeeper in Seattle's Woodland Zoo – participated in the Klondike gold rush in 1897–98 and collected snails and slugs during his travels on the Chilcoot Trail. In 1899, after his return to Seattle, Randolph wrote about his journey: "Packing 100 pounds over a pass 3,000 feet high did not tend to arouse my conchological ambition, but at each stop I prospected the dead leaves and sticks with varying success." C. Montague Cooke (1874–1948), a resident of Hawaii and associated with the Bishop Museum in Honolulu, spent part of the summer of 1922 on Vancouver Island, and there discovered a new species of land snail, now known as *Microphysula cookei*. Junius Henderson (1865–1937), curator of the University of Colorado Museum in

Boulder and best known for his important faunistic works on the non-marine molluscs of the western states, travelled to northern British Columbia, Alaska and the southern Yukon in 1925.

Perhaps the first resident naturalist in British Columbia to study land snails and slugs was Reverend George W. Taylor (1851–1912). He collected on Vancouver Island, contributed specimens to museums and wrote a number of short papers, including a remarkably complete checklist for Vancouver Island (1889, 1891a). Another early resident of British Columbia with interests in malacology was Abdiel W. Hanham (1851–1944), who collected specimens from near his home in Duncan on Vancouver Island. Another early collector was Charles Newcombe (1851–1924), who amassed a large collection of natural history and anthropological artifacts; the Royal BC Museum purchased the Newcombe collection in 1960.

The American malacologist Henry Augustus Pilsbry (1862–1952) most influenced the study of terrestrial molluscs in North America. Affiliated with the Philadelphia Academy of Sciences throughout his career, Pilsbry was regarded as the foremost authority on North American land snails and slugs, although he did not limit his study just to these. His contributions in this field remain unsurpassed, with his major accomplishment his monograph, *Land Mollusca of North America (north of Mexico)* published between 1939 and 1948.

Selected references: Baird 1863; Berry 1922; Drake 1963; Forsyth 1999a; Henderson 1927; La Rocque 1962; Lord 1866; Pilsbry and Cooke 1922; Randolph 1899; Taylor 1889, 1891a.

Important Literature

Important literature on the terrestrial molluscs of western North America: Bequaert and Miller 1973; Dall 1905; La Rocque 1953; Metcalf and Smartt 1997; Pilsbry 1939, 1940, 1946, 1948; Roth and Sadeghian 2003; Turgeon et al. 1998.

Books on terrestrial molluscs of Europe (including introduced and Holarctic species): Adam 1960; Backhuys 1975; Barker 1999; Cameron et al. 1983; Ellis 1969; Gittenberger et al. 1984; Hanna 1966; Kerney 1999; Kerney and Cameron 1979 (and subsequent editions in German and French – Kerney et al. 1983, Kerney and Cameron 2002); Likharev and Rammel'meier 1962.

CHECKLIST

This checklist includes all species of terrestrial snails and slugs known to live in British Columbia. Introduced species are marked with an asterisk. Extralimital or doubtful species, mentioned only briefly in the Species Accounts, do not appear in this list.

The classification of the Gastropoda, especially at higher levels, continues to undergo major changes as scientists rework the systematics of this diverse group of animals. The taxonomy used in the following list and species accounts mostly follows the classifications of some recent European authors (e.g., Falkner et al. 2000) and melds recent advances in the systematics of the Gastropoda with the long established classifications of Pilsbry (1936–48), Vaught (1989) and others. Important works dealing with gastropod systematics include Ponder and Lindberg 1996, Hausdorf 1998, Emberton et al. 1990, Wade et al. 2001, Yoon and Kim 2000, Dutra-Clarke et al. 2001, and Roth and Sadeghian 2003.

All land snails and slugs belong to the Phylum Mollusca and Class Gastropoda; all those described in this book belong to the Order Pulmonata. From there, I use the taxonomic categories of suborder, family and subfamily, and omit the intermediate categories (infraorder, etc.), which are unnecessary for most readers. The subgenus names (in parentheses after the genus name) show the relationships of species within the genus.

Phylum Mollusca – Class Gastropoda – Order Pulmonata

Suborder Acteophila
Family Carychiidae
Genus *Carychium* Müller, 1774
    Carychium minimum Müller, 1774*
    Carychium occidentale Pilsbry, 1891

Suborder Stylommatophora
Family Succineidae
  Subfamily Catinellinae
    Genus *Catinella* Pease, 1871
      Subgenus *Mediappendix* Pilsbry, 1948
        Catinella (Mediappendix) vermeta (Say, 1829)
  Subfamily Succineinae
    Genus *Oxyloma* Westerlund, 1885
      Oxyloma groenlandicum (Møller, 1842)
      Oxyloma hawkinsii (Baird, 1863)
      Oxyloma nuttallianum (I. Lea, 1841)
Family Cionellidae
  Genus *Cochlicopa* Risso, 1826
    Cochlicopa lubrica (Müller, 1774)
Family Lauriidae
  Genus *Lauria* J.E. Gray *in* Turton, 1830
    Subgenus *Lauria* s.s.
      Lauria (Lauria) cylindracea (da Costa, 1778)*
Family Valloniidae
  Subfamily Acanthinulinae
    Genus *Planogyra* Morse, 1864
      Planogyra clappi (Pilsbry, 1898)
    Genus *Zoogenetes* Morse, 1864
      Zoogenetes harpa (Say, 1824)
  Subfamily Valloniinae
    Genus *Vallonia* Risso, 1826
      Vallonia cyclophorella Sterki, 1892
      Vallonia excentrica Sterki *in* Pilsbry, 1893*
      Vallonia gracilicosta Reinhardt, 1883
      Vallonia pulchella (Müller, 1774)*
Family Pupillidae
  Subfamily Pupillinae
    Genus *Pupilla* Turton, 1831
      Subgenus *Pupilla* s.s.
        Pupilla (Pupilla) hebes (Ancey, 1881)
Family Vertiginidae
  Subfamily Gastrocoptinae

Genus *Gastrocopta* Wollaston, 1878
  Subgenus *Albinula* Sterki, 1892
    Gastrocopta (Albinula) holzingeri (Sterki, 1889)
Subfamily Truncatellininae
  Genus *Columella* Westerlund, 1878
    Columella columella (von Martens, 1830)
    Columella edentula (Draparnaud, 1805)
Subfamily Vertigininae
  Genus *Nearctula* Sterki, 1892
    Nearctula species
  Genus *Vertigo* Müller, 1774
    Subgenus *Vertigo* s.s.
      Vertigo (Vertigo) andrusiana Pilsbry, 1899
      Vertigo (Vertigo) arthuri von Martens, 1882
      Vertigo (Vertigo) columbiana Pilsbry & Vanatta, 1900
      Vertigo (Vertigo) cristata Sterki *in* Pilsbry, 1919
      Vertigo (Vertigo) elatior Sterki, 1893
      Vertigo (Vertigo) gouldii (A. Binney, 1843)
      Vertigo (Vertigo) modesta (Say, 1824)
      Vertigo (Vertigo) ovata Say, 1822
      Vertigo (Vertigo) species
Family Haplotrematidae
  Subfamily Haplotrematinae
    Genus *Ancotrema* H.B. Baker, 1931
      Ancotrema hybridum (Ancey, 1888)
      Ancotrema sportella (Gould, 1846)
    Genus *Haplotrema* Ancey, 1881
      Subgenus *Ancomena* H.B. Baker, 1931
        Haplotrema (Ancomena) vancouverense (I. Lea, 1839)
Family Testacellidae
  Genus *Testacella* Draparnaud, 1801
    Subgenus *Testacella* s.s.
      Testacella (Testacella) haliotidea Draparnaud, 1801*
Family Punctidae
  Genus *Paralaoma* Iredale, 1913
    Paralaoma servilis (Shuttleworth, 1852)
  Genus *Punctum* Morse, 1864
    Subgenus *Punctum* s.s.
      Punctum (Punctum) randolphii (Dall, 1895)
Family Discidae
  Genus *Anguispira* Morse, 1864
    Subgenus *Zonodiscus* Pilsbry, 1948
      Anguispira (Zonodiscus) kochi (Pfeiffer, 1821)
  Genus *Discus* Fitzinger, 1833
    Subgenus *Antediscus* H.B. Baker *in* Pilsbry, 1948

Discus (Antediscus) shimekii (Pilsbry, 1890)
Subgenus *Discus* s.s.
Discus (Discus) whitneyi (Newcomb, 1864)
Subgenus *Patula* Held, 1837
Discus (Patula) rotundatus (Müller, 1774)*
Family Oreohelicidae
Genus *Oreohelix* Pilsbry, 1904
Oreohelix strigosa (Gould, 1846)
Oreohelix subrudis (Reeve, 1854)
Family Pristilomatidae
Genus *Pristiloma* Ancey, 1887
Subgenus *Priscovitrea* H.B. Baker, 1931
Pristiloma (Priscovitrea) chersinella (Dall, 1866)
Pristiloma (Priscovitrea) johnsoni (Dall, 1895)
Subgenus *Pristiloma* s.s.
Pristiloma (Pristiloma) lansingi (Bland, 1875)
Pristiloma (Pristiloma) stearnsii (Bland, 1875)
Subgenus *Pristinopsis* H.B. Baker, 1931
Pristiloma (Pristinopsis) arcticum (Lehnert, 1884)
Genus *Vitrea* Fitzinger, 1833
Subgenus *Crystallus* Lowe 1855
Vitrea (Crystallus) contracta (Westerlund, 1871)*
Family Euconulidae
Subfamily Euconulinae
Genus *Euconulus* Reinhardt, 1833
Subgenus *Euconulus* s.s.
Euconulus (Euconulus) fulvus (Müller, 1774)
Euconulus (Euconulus) praticola (Reinhardt, 1883)
Family Gastrodontidae
Genus *Striatura* Morse, 1864
Subgenus *Pseudohyalina* Morse, 1864
Striatura (Pseudohyalina) pugetensis (Dall, 1895)
Genus *Zonitoides* Lehmann, 1862
Subgenus *Zonitoides* s.s.
Zonitoides (Zonitoides) arboreus (Say, 1816)
Zonitoides (Zonitoides) nitidus (Müller, 1774)
Family Daudebardiidae
Subfamily Godwiniinae
Genus *Aegopinella* Lindholm, 1927
Aegopinella nitidula (Draparnaud, 1805)*
Genus *Nesovitrea* C.M. Cooke, 1921
Subgenus *Perpolita* H.B. Baker, 1928
Nesovitrea (Perpolita) binneyana (Morse, 1864)
Nesovitrea (Perpolita) electrina (Gould, 1841)
Subfamily Oxychilinae

Genus *Oxychilus* Fitzinger, 1833
  Subgenus *Oxychilus* s.s.
    Oxychilus (Oxychilus) alliarius (J.S. Miller, 1822)*
    Oxychilus (Oxychilus) cellarius (Müller, 1774)*
    Oxychilus (Oxychilus) draparnaudi (Beck, 1837)*
Family Vitrinidae
Subfamily Vitrininae
  Genus *Vitrina* Draparnaud, 1801
    Vitrina pellucida (Müller, 1774)
Family Boettgerillidae
  Genus *Boettgerilla* Simroth, 1910
    Boettgerilla pallens Simroth, 1912*
Family Limacidae
Subfamily Limacinae
  Genus *Limacus* Lehmann, 1864
    Limacus flavus (Linnaeus, 1758)*
  Genus *Limax* Linnaeus, 1758
    Limax maximus Linnaeus, 1758*
  Genus *Lehmannia* Heynemann, 1863
    Lehmannia valentiana (Férussac *in* Férussac & Deshayes, 1822)*
Family Agriolimacidae
Subfamily Agriolimacinae
  Genus *Deroceras* Rafinesque, 1820
    Subgenus *Deroceras* s.s.
      Deroceras (Deroceras) hesperium Pilsbry, 1944
      Deroceras (Deroceras) laeve (Müller, 1774)
      Deroceras (Deroceras) panormitanum (Lessona & Pollonera, 1882)*
      Deroceras (Deroceras) reticulatum (Müller, 1774)*
Family Arionidae
Subfamily Ariolimacinae
  Genus *Ariolimax* Mørch, 1860
    Subgenus *Ariolimax* s.s.
      Ariolimax (Ariolimax) columbianus (Gould *in* A. Binney, 1851)
  Genus *Magnipelta* Pilsbry, 1953
    Magnipelta mycophaga Pilsbry, 1953
Subfamily Arioninae
  Genus *Arion* Férussac, 1819
    Subgenus *Arion* s.s.
      Arion (Arion) rufus (Linnaeus, 1758)*
    Subgenus *Carinarion* Hesse, 1926
      Arion (Carinarion) circumscriptus Johnston, 1828*
      Arion (Carinarion) silvaticus Lohmander, 1937*
    Subgenus *Kobeltia* Simroth, 1873
      Arion (Kobeltia) distinctus Mabille, 1868*
      Arion (Kobeltia) intermedius Normand, 1852*

Subgenus *Mesarion* Hesse, 1926
Arion (Mesarion) subfuscus (Draparnaud, 1805)*
Subfamily Binneyinae
Genus *Hemphillia* Bland & Binney, 1872
Hemphillia camelus Pilsbry & Vanatta, 1897
Hemphillia dromedarius Branson, 1972
Hemphillia glandulosa Bland & W.G. Binney, 1872
Subfamily Anadeninae
Genus *Prophysaon* Bland & W.G. Binney, 1873
Subgenus *Mimetarion* Pilsbry, 1948
Prophysaon (Mimetarion) vanattae Pilsbry, 1948
Subgenus *Prophysaon* s.s.
Prophysaon (Prophysaon) andersonii (J.G. Cooper, 1872)
Prophysaon (Prophysaon) coeruleum Cockerell, 1890
Prophysaon (Prophysaon) foliolatum (Gould *in* A. Binney, 1851)

Family Polygyridae
Subfamily Polygyrinae
Genus *Allogona* Pilsbry, 1939
Subgenus *Dysmedoma* Pilsbry, 1939
Allogona (Dysmedoma) ptychophora (A.D. Brown, 1870)
Allogona (Dysmedoma) townsendiana (I. Lea, 1839)
Genus *Cryptomastix* Pilsbry, 1939
Subgenus *Cryptomastix* s.s.
Cryptomastix (Cryptomastix) devia (Gould, 1846)
Cryptomastix (Cryptomastix) mullani (Bland & J.G. Cooper, 1881)
Subgenus *Micranepsia* Pilsbry, 1940
Cryptomastix (Micranepsia) germana (Gould *in* A. Binney, 1851)
Genus *Vespericola* Pilsbry, 1939
Vespericola columbianus (I. Lea, 1839)

Family Thysanophoridae
Genus *Microphysula* Pilsbry, 1926
Microphysula cookei (Pilsbry, 1922)
Microphysula ingersollii (Bland, 1875)

Family Bradybaenidae
Genus *Monadenia* Pilsbry, 1895
Subgenus *Monadenia* s.s.
Monadenia (Monadenia) fidelis (J.E. Gray, 1834)

Family Helicidae
Subfamily Helicinae
Genus *Cepaea* Held, 1837
Subgenus *Cepaea* s.s.
Cepaea (Cepaea) nemoralis (Linnaeus, 1758)*
Genus *Cornu* Born, 1778
Cornu aspersum (Müller, 1774)*

IDENTIFICATION AND KEYS TO GENERA

Identifying Snails and Slugs

Many of the species in this book are very small, so a good hand lens of 10–20× magnification or a low-power microscope are indispensable; even the largest snails and slugs can be identified more easily, or at least the fine details better noticed, with a hand lens. To measure specimens, use a set of callipers or a ruler showing millimetres. If you have a microscope, then a scale in the eyepiece is ideal for measuring minute species.

Most snails can be identified by shell characters, but it is especially important to have fresh, full-grown shells. Worn or faded shells are often unsuitable for identification. Juvenile shells can be a problem as well since they may appear different from adults and lack diagnostic characters, such as denticles in the aperture. With experience, your recognition of adults and corresponding juveniles will become easier, and with practice on fresh material, old worn shells may be identifiable.

The main external characters used for identifying slug genera in this key are the size and shape of the animal, relative position of the pneumostome (figure 21), texture of the mantle, extent to which the keel on the tail is developed, and the presence and placement of an external shell. The colour and texture of the skin and the stickiness and colour of the mucus can help in recognizing slug species. Skin colour fades and discolours quickly in alcohol-preserved specimens, so live animals or good photographs are best to work with when looking for pigmentation. In some cases, dissection is unavoidable to confirm species identification; the internal anatomy (particularly of the distal genitalia, but also of the digestive system) is important for distinguishing a species.

Variation in the colour and form of animals and their shells is common, and should be expected. (See Goodfriend 1986 for a review of variation in land-snail shells and its causes.)

Key to Snails

This key includes all of British Columbia's terrestrial gastropod genera that have an obviously coiled shell into which the animal can retract. The key is for fully grown specimens. Shells of juveniles usually have a different form than those of full-grown snails and usually lack the apertural denticles or developed lip present in adults of some genera.

1a Shell height greater than width 2
1b Shell height not greater than width 10
2a Aperture with one or more denticles 3
2b No apertural denticles 6
3a Shell whitish or colourless 4
3b Shell yellowish, brown, reddish brown 5
4a Shell more or less cylindrical .... *Gastrocopta*, p. 53
4b Shell spindle-shaped to elongate-tapering ..... *Carychium*, p. 36
5a Angular lamella prominent, usually connected to the apertural lip by a narrow, white callus; no other denticles are present on the parietal wall; palatal denticles absent (although lip is considerably thickened and swollen medially) .... *Lauria*, p. 42
5b Angular lamella small or absent, but other denticles are present on the parietal wall; two or more palatal denticles .... *Vertigo, Nearctula*, p. 58
6a Shell subcylindrical 7
6b Shell clearly not subcylindrical 9
7a Shell surface very smooth and with a high gloss; umbilicus absent .... *Cochlicopa*, p. 41
7b Shell surface not exceptionally smooth and glossy; umbilicus present 8
8a Apertural lip narrowly flared and slightly thickened; crest present .... *Pupilla*, p. 51

8b	Apertural lip neither flared nor thickened; crest absent .... *Columella*, p. 54
9a	Shell with well-spaced, *lamellar* axial ribs ...... *Zoogenetes*, p. 45
9b	Shell nearly smooth, with incremental striae only* .... *Oxyloma, Catinella*, p. 39
10a	Umbilicus closed or tiny (a minute pit)................................... 11
10b	Umbilicus open 16
11a	Shell width greater than 15 mm, spiral colour bands often present... 12
11b	Shell width less than 9 mm, without colour bands............... 13
12a	Aperture height less than aperture width; typically banded (bands continuous, seldom interrupted by axial streaks) or solid coloured .... *Cepaea*, p. 160
12b	Aperture height about equal to its width; banded, with bands usually interrupted by pale axial streaks............. *Cornu*, p. 162
13a	Parietal denticle present; periostracum hairy (but may be worn off).. *Cryptomastix*, p. 151
13b	Parietal denticle absent; periostracum hairless 14
14a	Aperture broadly and deeply rounded; about 3 whorls .... *Vitrina*, p. 109
14b	Aperture narrowly crescent-shaped; 3½ or more whorls 15
15a	Shell with minute umbilicus and exceedingly fine, close, threads and spiral striae.................................... *Euconulus*, p. 94
15b	Shell without an umbilicus and smooth, axially ribbed or with incremental striae .... *Pristiloma*, p. 87
16a	Apertural lip strongly flared or recurved 17
16b	Aperture lip not strongly flared or recurved......................... 19
17a	Parietal denticle present............................. *Cryptomastix*, p. 151
17b	Parietal denticle absent 18
18a	Periostracum not hairy; shell with strong axial threads, ribs or wrinkles, sometimes with dimpled sculpture; minute regular spiral striae present on uneroded areas of the shell .... *Allogona*, p. 148
18b	Periostracum hairy (but occasionally worn off); shell with incremental striae only .... *Vespericola*, p. 155

---

* Do not confuse these succineid genera with semi-aquatic snails of the family Lymnaeidae, and in particular with *Pseudosuccinea columella* (Say, 1817). Lymnaeids have 4–6 whorls and the columella is strongly twisted. Lymnaeids also lack ocular tentacles and their eyes are at the bases of the sensory tentacles, which are rudimentary in succineids.

19a Shell less than 8 mm wide 26
19b Shell at least 8 mm wide 20
20a Shell usually has spiral colour bands (but not always) 21
20b Shell without spiral colour bands... 23
21a Shell usually chestnut brown with a narrow, pale yellowish band at the periphery, a narrow dark brown band above this, and a dark brown base (yellow shells with pale banding are occasional) .... *Monadenia*, p. 159
21b Shell with bands unlike those above, or without bands....... 22
22a Shell brown with two obscure dark bands (one at the periphery, the just above it) that usually delimit a paler band .... *Anguispira*, p. 79
22b Usually whitish or greyish, often stained with brown, and usually with brown spiral bands, but not as above .... *Oreohelix*, p. 84
23a Apertural lip slightly thickened and concave or straightened at the shoulder in apertural view 24
23b Apertural lip thin and not as above.. 25
24a Spiral striae intersect axial ribs to form beaded sculpture (at least on the whorls of the spire) .... *Ancotrema*, p. 70
24b Shell surface not beaded; spiral striae very fine and axial ribs not developed .... *Haplotrema*, p. 72
25a Shell with fine, faint spiral striae and a waxy sheen; umbilicus about a quarter the width of the shell .... *Aegopinella*, p. 101
25b Shell without spiral striae and more-or-less glossy; umbilicus about a sixth the width of the shell .... *Oxychilus*, p. 104
26a Shell whitish, greyish-white or colourless 27
26b Shell yellowish or brownish (caution: long-dead, partially decayed shells may be whitish).. 29
27a Shell with more-or-less regularly and widely spaced lamellar axial ribs; or, if ribs not developed, then apertural lip thick .... *Vallonia*, p. 46
27b Shell without lamellar ribs; apertural lip thin........................ 28
28a Umbilicus smaller, about a sixth of the shell width; spiral striae absent.. *Vitrea*, p. 92
28b Umbilicus larger, about a quarter of the shell width; microscopic spiral striae present......................... *Microphysula*, p. 156
29a Umbilicus at least a third of the shell width........................... 30
29b Umbilicus less than a third of the shell width........................ 32
30a Shell less than 1.8 mm 31

30b Shell more than 1.8 mm, up to 7.5 mm.................. *Discus*, p. 80

31a Shell dark brown; axial ribs delicate, lamellar and distantly
     spaced .... *Planogyra*, p. 44

31b Shell paler, yellowish or yellowish brown; axial ribs fine,
     closely spaced and rounded rather than lamellar (and nearly
     equal in size to the spiral sculpture)................... *Striatura*, p. 97

32a Shell with widely and more-or-less regularly spaced lamellar
     axial ribs (sometimes worn off); protoconch initially smooth,
     then with widely spaced spiral threads........... *Paralaoma*, p. 75

32b Shell smooth or with axial ribs, riblets or striae densely
     packed and never lamellar (except at high magnification).. 33

33a Shell tiny, width to 1.8 mm (3½ – 4½ whorls)... *Punctum*, p. 77

33b Shell larger 34

34a Aperture narrowly crescent-shaped .... *Pristiloma*, p. 87

34b Aperture broader.. 35

35a Major axial sculpture consists of more-or-less regularly-
     spaced indented groves.................................... *Nesovitrea*, p. 102

35b Shell smooth or axially striate (but not as above)................. 36

36a Whorls rapidly enlarge in width: last whorl at least twice the
     width of the penultimate whorl one-half turn back from
     aperture (in apical view).................................... *Oxychilus*, p. 104

36b Whorls less rapidly enlarge in width: at one-half coil back
     from aperture, last whorl less than twice the width of the
     penultimate whorl... *Zonitoides*, p. 98

Key to Slugs

This key is to the genera of slug-like terrestrial molluscs as they occur in British Columbia. Most have no external shell; those that do have an external shell cannot retract into it.

1a   Shell completely or partially exposed, either plate-like and
      mostly embedded within the mantle near the middle or front
      of the animal, or ear-shaped and at the tip of the tail............. 2

1b   Shell not exposed 3

2a   Shell ear-shaped at the tip of the tail .... *Testacella*, p. 74

2b   Shell plate-like, mostly embedded in a well-defined visceral
      hump near the longitudinal middle of the animal
      .... *Hemphillia*, p. 136

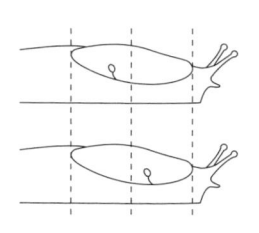

Figure 21. Relative position of the pneumostome.

3a   Narrow, wormlike body, pale greyish .... *Boettgerilla*, p. 110
3b   Body not as above 4
4a   Pneumostome at or in front of the midline of the mantle on the right side (figure 21) 5
4b   Pneumostome behind the midline of the mantle on the right side 6
5a   Tail often having an oblique constriction (and sometimes a darker line on the sole of the foot) that marks the site of autotomy; caudal mucus pore absent; penis present .... *Prophysaon*, p. 140
5b   Tail never without an oblique constriction; caudal mucus pore present; penis absent .... *Arion*, p. 126
6a   Mantle smooth or granular, without concentric folds or ridges 7
6b   Mantle with a pattern of concentric folds or fingerprint-like ridges 8
7a   Mantle very large, covering most of the dorsal surface of the animal; dorsal keel absent .... *Magnipelta*, p. 125
7b   Mantle not exceptionally large; well developed dorsal keel .... *Ariolimax*, p. 123
8a   Tail gradually tapering to a point; mantle ridges centred on the midline; dorsal keel long (about a quarter the length of the body) 9
8b   Tail obliquely truncated; mantle ridges centred on right side above the pneumostome; dorsal keel short .... *Deroceras*, p. 116
9a   Mantle usually with distinct dark lateral bands (but sometimes faint) .... *Lehmannia*, p. 115
9b   Mantle solid coloured, spotted or marbled, but without bands 10
10a Body yellowish with grey mottling; tentacles bluish; body mucus yellowish .... *Limacus*, p. 112
10b Body greyish with darker mottling; tentacles reddish brown; body mucus colourless .... *Limax*, p. 113

SPECIES ACCOUNTS

This handbook describes 92 species of land snails and slugs known to occur in British Columbia. I have arranged the families systematically, and the subfamilies, genera and species within each family alphabetically. Each species account for confirmed species (except for those belonging to the family Succineidae) is divided into the following sections:

Names and Synonyms: Most of the common names are from *Common and Scientific Names of Aquatic Invertebrates from the United States and Canada: Mollusks, Second Edition* (Turgeon et al. 1998). For the few species not included in that edition, I took common names from other publications or named them myself, following the principles outlined in Turgeon et al. For synonyms, I list only the original genus-species combinations, and for most species only the western North American synonyms. If a species does not appear in Pilsbry's *Land Mollusca of North America (North of Mexico)* (1939–48), or if there are recent taxonomic works on a species, then I present a more complete list of synonyms.

Description: A brief account of the most important features required for identification. The shell measurement is the greater of either the height or width and represents the maximum adult size. Slug measurements are of fully extended living adult animals. Whorls are counted to the nearest quarter whorl (see figure 19). I include genital characters only where they are needed for identification, but cite key references that describe or illustrate genitalia. Following the main description, I compare the external appearance of the living animals with similar species and point out distinguishing characters.

Distribution: The serious lack of records makes distribution maps pointless. Instead, I describe the range of the species worldwide and in British Columbia.

Natural History: This summary of habitat, diet, behaviour and reproduction is short because we know little about most species in the province.

Etymology: A summary of the definition and derivation of the scientific name.

Remarks: Information not given elsewhere appears here, including important notes on recent taxonomic changes and conservation status. I also list any recognized subspecies here.

Selected References: A list of important published literature on the species. Works published before Pilsbry's *Land Mollusca of North America* (1939–48) are included only if they contain more information than Pilsbry does in his summary of them.

Family Carychiidae

Carychium minimum **Müller, 1774**   **Herald Thorn**

Description
Shell 1.6–2.0 mm high, spindle-shaped and translucent-white or colourless, with weak incremental striae. Apertural lip thickened and slightly flared and with a central palatal denticle in fully developed shells; lip not straight when viewed from the side. Aperture has a prominent parietal denticle and a strongly developed columellar denticle (developed internally on the columella as spiral folds). Animal white with black eye spots behind the base of a single pair of tentacles. Genitalia: unknown.

Similar to *Carychium occidentale*, which is larger and thinner, has a thinner apertural lip and more weakly developed columellar and palatal denticles.

Distribution
Europe and Siberia. Introduced to southern Vancouver Island and San Francisco, California.

Natural History

This species requires extremely moist conditions. In Europe it lives in marshes, swamps and wet woods under rotten wood and stones, and in leaf litter and moss. In North America it is found in wet places in gardens and plant nurseries.

The life history of *Carychium minimum* is unknown, but a close relative, *C. tridentatum*, can reproduce uniparentally, lives from a few months to about 18 months, lays single eggs and eats leaf litter.

Etymology

Carychium: "a shepherd's horn" or "ancient, curved musical instrument"; *minimum*: "smallest".

Selected References

Adam 1960, Bank and Gittenberger 1985, Bulman 1990, Kerney 1999, Kerney and Cameron 1979, Likharev and Rammel'meier 1962, Morton 1955, Roth 1982a, Watson and Verdcourt 1953.

Carychium occidentale Pilsbry, 1891   Western Thorn
Synonym: *Carychium magnificum* Hanna, 1923.

Description

Shell 2.0–2.7 mm high, spindle-shaped but with a strongly tapering spire, translucent-white or colourless, with weak incremental striae. Apertural lip flared but not substantially thickened, and almost

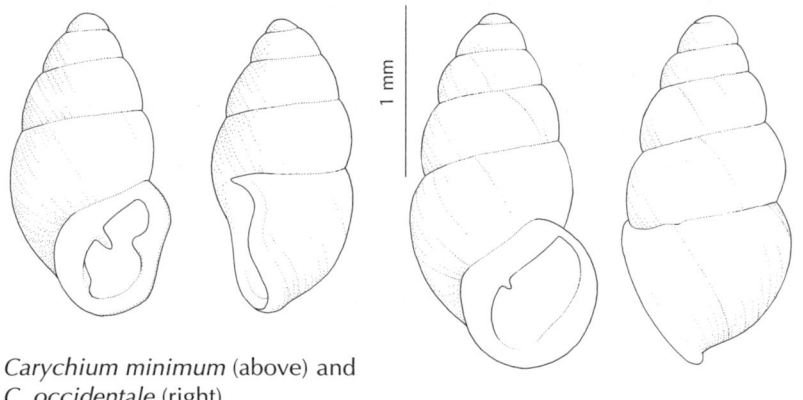

Carychium minimum (above) and
C. occidentale (right).

straight when viewed from the side. Aperture has a prominent parietal denticle (developed internally as a spiral lamella) and sometimes a weakly developed columellar denticle. Palatal denticle scarcely evident or absent. Animal white with black eye spots. Genitalia: unknown.

See comparison under *Carychium minimum*.

Distribution
Northern Vancouver Island, British Columbia, to Mendocino County, California, east through northern Washington to Idaho. In B.C., *Carychium occidentale* is known only from the west side of the Coast and Cascade ranges, but should be expected in the southeastern part of the province.

Natural History
This species lives in leaf litter in rich, relatively undisturbed low-elevation forests that are usually dominated by Bigleaf Maple.

Etymology
Carychium: "a shepherd's horn" or "ancient, curved musical instrument"; *occidentale*: "western".

Selected References
Branson 1977, Branson and Branson 1984, Cameron 1986, Frest and Johannes 2001, Hanna 1923, Pilsbry 1948.

Unconfirmed species of Carychiidae

Carychium exile H.C. Lea, 1842   Ice Thorn
Hanna (1923) reported this species from Union Bay on Vancouver Island, as *C. exile canadense* Clapp, 1906, but this record was probably mislabelled. *C. exile* occurs in central and eastern North America and has more distinct, almost rib-like, axial sculpture than either *C. minimum* or *C. occidentale*. It is also smaller than *C. occidentale* and the apertural lip is clearly thickened and slightly flared.
Selected references: Nekola and Barthel 2002, Pilsbry 1948.

Family Succineidae

Shells of these wetland snails are so featureless and vary to such a degree that they cannot be used to identify a species; reproductive anatomy is far more useful in identification. Except for the study by Stuart Harris and Leslie Hubricht (1982), there are no recent comparisons of material from British Columbia to named taxa outside the province. Given the difficulties encountered with the group, and that the number of species of succineids in British Columbia is uncertain, I only list the species that we know or suspect live here; it is likely that exotic species may also be present.

Catinella (Mediappendix) gabbii (Tryon, 1866)
Riblet Ambersnail
Found in eastern Washington, Oregon, and Idaho to Utah and northeastern California; possibly in British Columbia.
Selected references: Pilsbry 1948, Roth and Sadeghian 2003.

Catinella (Mediappendix) rehderi (Pilsbry, 1948)
Chrome Ambersnail
Recorded in Montana and Washington to southern California and northwestern Baja California, Mexico; possibly in B.C.
Selected references: Pilsbry 1948, Roth and Sadeghian 2003.

Catinella (Mediappendix) vermeta (Say, 1829)
Suboval Ambersnail
Synonym: *Succinea avara* (Say, 1824).
Recorded from much of North America, including B.C., but this may be a complex of several species.
Selected references: Franzen 1982, Pilsbry 1948.

Oxyloma groenlandicum (Møller, 1842) Ruddy Ambersnail
Synonyms: *Succinea groenlandica* Møller, 1842; ?*S. verrilli* Bland, 1865.
This species inhabits subarctic portions of the Yukon Territory to south-central British Columbia, and east to eastern Canada,

Greenland and Iceland. It is possibly the same species, which in Europe has gone by the name *Oxyloma pfeifferi* (Rossmässler, 1835) or *O. elegans* (Risso, 1826).
Selected references: Harris and Hubricht 1982, Faulkner et al. 2002, Pilsbry 1948.

Oxyloma hawkinsii (Baird, 1863)　　Boundary Ambersnail
Found in southern British Columbia, Alberta, Washington and Idaho.
Selected references: Frest and Johannes 2001, Harris and Hubricht 1982, Pilsbry 1948.

Oxyloma nuttallianum (I. Lea, 1841)　　Oblique Ambersnail
Synonyms: ?*Succinea oregonensis* of authors, not I. Lea, 1841; *S. rusticana* of authors, not Gould, 1856.
This species ranges from British Columbia to Baja California, and is believed to be widespread in western North America.
Selected references: Franzen 1985, Harris and Hubricht 1982, Pilsbry 1948, Smith et al. 1990.

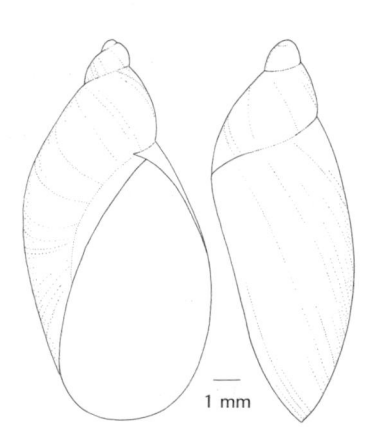

Oxyloma nuttallianum.

Family Cionellidae

Cochlicopa lubrica (Müller, 1774)   **Glossy Pillar**
Synonym: *Helix lubrica* Müller, 1774.

Description
Shell 4.4–6.5 mm high, spindle-shaped, translucent, light brown, glossy and almost smooth. Spire long with a bluntly rounded apex. Apertural lip thickened within by an opaque whitish or pinkish rib. Umbilicus absent. Animal dark grey to bluish-black, paler on the sides. Genitalia: Quick 1954, Outeiro et al. 1990, Armbruster 1994.

Full-grown shells are easily distinguished from all other land snails in our area by the combination of their high gloss, spindle-shape and thickened apertural lip.

Distribution
Throughout much of Europe, North Africa, northern Asia and North America south to Mexico; it is introduced to New Zealand, Venezuela, southern Africa and several islands in the North and South Atlantic. *Cochlicopa lubrica* likely occurs throughout British Columbia.

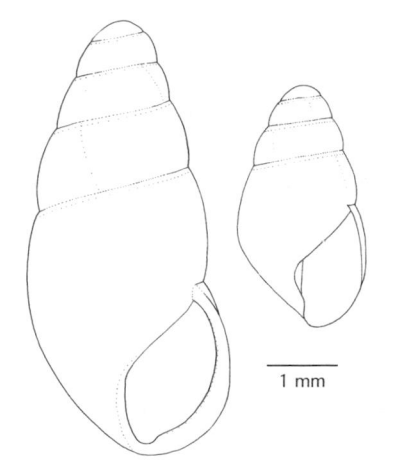

1 mm

Adult (left) and juvenile.

Natural History
Cochlicopa lubrica is strongly synanthropic and appears mostly at disturbed sites such as roadsides, gardens and pastures. It is uncommon in natural areas, such as seaside open forests and dunes in the Queen Charlotte Islands, and has not been found in dense forests. Most urban populations are probably introduced. It is gregarious under rocks, dead wood, leaf litter, vegetation and debris. Some people have recorded great numbers of these snails on a patch of ground, a concrete walk, wall or foundation – one such congregation was seen at Okanagan Centre. These snails eat both dead and live plant material. During aestivation, they seal their aperture with

a transparent, colourless epiphragm. This and other species in the genus are believed to reproduce predominantly by self-fertilization.

Etymology
Cochlicopa: "conical snail"; *lubrica:* "slippery", suggesting the difficulty often encountered when handling the smooth shell.

Remarks
The recognition of species in this genus is complicated. Recent genetic studies of European *Cochlicopa* revealed that shell similarities and differences among "species" do not correspond to genetic groups. On the other hand, Yaroslav Starobogatov inappropriately used shell measurements and other slight differences in shells to justify a large number of Eurasian *Cochlicopa* species.

Selected References
Armbruster 1994, 1995, 1997, 2001; Armbruster and Bernhard 2000; Armbruster and Schlegel 1994; Barker 1999; Bequaert and Miller 1973; Caesar 1946; Gittenberger et al. 1984; Giusti and Manganelli 1992; Hudec 1960; Kerney 1999; Kerney and Cameron 1979; Pilsbry 1948; Roscoe 1962; Roth and Pearce 1984; Starobogatov 1996.

Family Lauriidae

Lauria cylindracea (da Costa, 1778)   Chrysalis Snail
Synonym: *Turbo cylindracea* da Costa, 1778.

Description
Shell 4.1 mm high, subcylindrical to subovate, pale brown, with fine incremental striae and 6 whorls. Aperture has a pointed, narrow parietal denticle that is usually connected to the lip by a callus. Apertural lip white, and expanded and recurved to form a broad flat face. The animal has a whitish foot and grey head and tentacles. Genitalia: Barker 1999.

The general form of the shell and development of the apertural teeth vary greatly within populations. Occasional specimens lack apertural denticles, whereas other specimens may have three strongly developed denticles (angular, columellar and sub-columellar) as well as a conspicuously thickened, almost dentate apertural lip. Juvenile shells have a series of palatal denticles visible through the basal shell wall, a distinct denticle on the columella, and a narrow, spiral ridge on the parietal wall that extends deep into the last whorl.

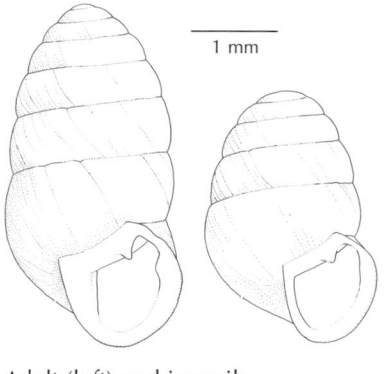

Adult (left) and juvenile.

Lauria cylindracea differs from *Pupilla hebes* and *P. muscorum* by the absence of a crest behind the lip. The apertural dentition in adult shells is also different, and *Pupilla* juveniles never have any trace of denticles. Very young individuals superficially resemble *Punctum randolphii* in shape but differ by the presence of denticles.

Distribution
Western Europe and the Mediterranean region east to the Caucasus and Asia Minor. Introduced to New Zealand, South Africa, several Atlantic islands, Reunion Island and British Columbia. This species is widespread in Victoria and Vancouver and occasionally found elsewhere on southern Vancouver Island and the Gulf Islands, and in the Chilliwack area.

Natural History
In British Columbia, *Lauria cylindracea* is synanthropic, living in dry to moist gardens, parks and waste ground, especially where garden waste accumulates. It is common under plants, fallen leaves, wood and stones, in old, untended gardens, and in disturbed sites under English Ivy and other exotic plants. In Europe, it lives in woods, hedgerows, grasslands and open rocky areas or rock walls. At times of drought, a seal of dried mucus cements the broad lip surrounding the edge of the aperture to a surface.

Lauria cylindracea is ovoviviparous and during drought may delay the release of young. It is relatively long-lived for a small snail, reaching five years of age in Israel.

Etymology
Lauria: perhaps named for the bay tree (*Laurus*); *cylindracea:* "cylindrical".

Selected References
Arad et al. 1998, Barker 1999, Forsyth 1999b, Gittenberger et al. 1984, Heller et al. 1997, Kerney 1999, Kerney and Cameron 1979.

Family Valloniidae

Planogyra clappi (Pilsbry, 1898)   Western Flat-whorl
Synonym: *Punctum clappi* Pilsbry, 1898.

Description
Shell 1.7–2.2 mm wide, flattened-heliciform and brown with delicate, widely spaced lamellar axial ribs having much finer incremental and spiral striae between. The lamellar ribs, which consist mostly of periostracum, are easily worn off and may be completely lacking or present only on the most recent part of the last whorl. Whorls 3 to 3½; spire nearly flat. Apertural lip thin and not expanded. Umbilicus large, about a third the width of the shell. Genitalia: unknown.

1 mm

Similar to *Paralaoma servilis,* which has a higher spire, a narrower umbilicus and is slightly larger. Species of *Vallonia* have white or greyish white shells and are larger. Other minute snails to which this species should be compared are *Striatura pugetensis* and *Punctum randolphii.*

Distribution
Queen Charlotte Islands and Vancouver Island, British Columbia, to Mendocino County, California, and in northern Idaho. *Planogyra*

clappi is widespread and moderately common on the west side of the Coast Range in B.C., but is so far unrecorded east of the Coast Mountains.

Natural History
This species is sporadically common in moist, rich coastal forests, where it lives in the leaf litter.

Etymology
Planogyra: "flat whorl"; *clappi:* after Dr George H. Clapp, an American conchologist.

Selected References
Frest and Johannes 2001, Pilsbry 1948, Roth 1985, Solem 1977b.

Zoogenetes harpa (Say, 1824)   **Boreal Top**
Synonym: *Helix harpa* Say 1824.

Description
Shell 3.3 mm high, conical-ovate and dark brown with widely spaced, lamellar axial ribs and finer incremental striae. About 4 whorls. Apertural lip thin and unexpanded. Umbilicus narrow. The animal is grey with darker ocular tentacles, a pale foot and a dark-grey mantle speckled with white. The foot has a scalloped edge. Genitalia: Pilsbry 1948.

No other species in this book are likely to be confused with *Zoogenetes harpa.*

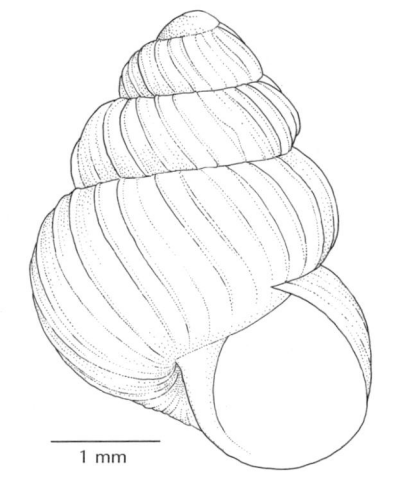

1 mm

Distribution
Northern Europe and northern Asia; in North America, throughout much of Canada and the northern United States south to Colorado.

In British Columbia, *Zoogenetes harpa* occurs sporadically north of the northern central interior and south along the Rockies.

Natural History
In British Columbia, *Zoogenetes harpa* lives in moist deciduous, coniferous and mixed-wood forests, in leaf litter, on vegetation, and under rocks and dead wood. It is ovoviviparous and possibly aphallic.

Etymology
Zoogenetes: "animal-birth", inferring live-born young; *harpa:* "a harp", a reference to the axial ribs, like the strings on a harp.

Selected References
Hubricht 1985, Kerney and Cameron 1979, Likharev and Rammel'meier 1962, Pilsbry 1948, Steenberg 1925.

Key to species of *Vallonia*

1a Shell has more-or-less regularly spaced axial ribs; last whorl clearly descending at the aperture 2
1b Shell without regularly spaced axial ribs; last whorl not descending at the aperture 3
2a Lip flared but not thickened inside by a callus
.... *Vallonia cyclophorella*
2b Lip at least slightly thickened inside by a callus. *V. gracilicosta*
3a Last whorl near the aperture distinctly increasing in width much more than earlier in the shell; outline of the shell ovate; umbilicus elongate, elliptically spiral; apertural lip gradually expanded, not at a right angle..................................*V. excentrica*
3b Last whorl near the aperture increasing in width only a little more than earlier; outline of the shell almost circular; umbilicus regularly spiral; apertural lip abruptly flared outward at approximately a right angle ....*V. pulchella*

Vallonia cyclophorella Sterki, 1892   Silky Vallonia
Synonym: *Vallonia cyclophorella septuagentaria* Pilsbry & Ferriss, 1918.

Description
Shell 3.4 mm wide, flattened-heliciform, translucent-white or greyish to brownish yellow, with regular, blade-like axial ribs and finer incremental striae between them. The last whorl descends at the aperture, which is ovate and clearly wider than its height. Apertural lip abruptly flared outward but not thickened within by a callus rib. Genitalia: unknown.

Similar to *Vallonia gracilicosta*, which as an adult has a thickened, opaque-white lip rib.

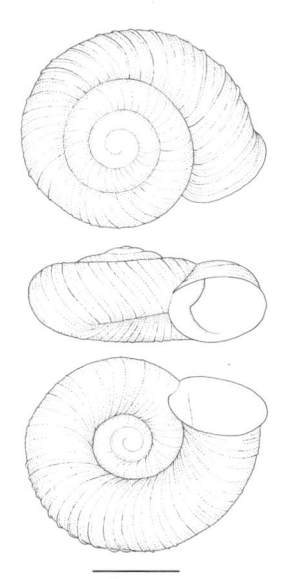

1 mm

Distribution
Pyramid Island, Alaska; southern British Columbia to west-central Alberta, south through Washington, Oregon, Idaho to southern California, Arizona, Colorado and New Mexico. In British Columbia, *Vallonia cyclophorella* is recorded from Kamloops, Lytton, Ashnola River and Skookumchuck. In Washington, it occurs only east of the Cascade Range; Branley Branson's record from the Olympic Peninsula is erroneous and based on *Paralaoma servilis*.

Natural History
Vallonia cyclophorella lives in leaf litter and grass and under dead wood on bunchgrass slopes, in open, dry forests and rocky areas.

Etymology
Vallonia: after the Roman goddess of the valleys; *cyclophorella*: diminutive of "ring-bearer".

Selected References
Bequaert and Miller 1973, Blood 1963, Branson 1977, Gerber 1996, Pilsbry 1948.

Vallonia excentrica Sterki *in* Pilsbry, 1893

Iroquois Vallonia

Description

Shell 2.2 mm wide, flattened-heliciform, with 3 to 3½ shiny, translucent-white whorls marked by fine, irregular axial striae and occasional low wrinkles. In apical view, the final portion of the last whorl enlarges rapidly, giving the shell an elongate outline. The last whorl does not descend before the aperture. Apertural lip broad, flat and conspicuously thickened within by a rib-like callus. Genitalia: Barker 1999.

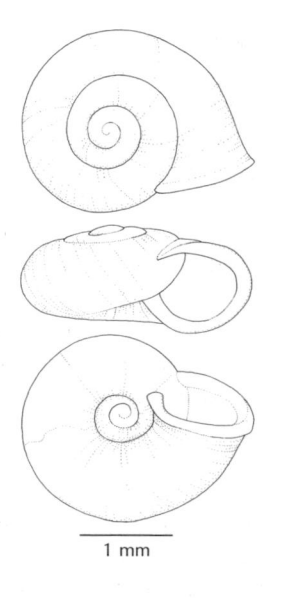

1 mm

Vallonia excentrica is similar to *V. pulchella*, but it is generally smaller, its last whorl is broadly expanded in its last quarter, its apertural lip flares more gradually, and the shell and umbilicus are more elliptical.

Distribution

Native to Europe, northern Asia and eastern and central North America; introduced to many places worldwide, including the U.S. Pacific coast states, and southern and central British Columbia.

Natural History

Vallonia excentrica is strongly synanthropic and common along roadsides and fields, and in lawns and gardens, where it lives under concrete, rocks, logs, debris and vegetation. It is often sympatric with *V. pulchella. V. excentrica* can be euphallic or aphallic. This snail lays single eggs.

Etymology

Vallonia: after the Roman goddess of valleys; *excentrica*: "off-centre", derived from off-centre appearance of the umbilicus.

Remarks

Recent research into the genetics of this taxon suggests that *Vallonia excentrica* is merely a loose arrangement of similar shell types.

Selected References

Forsyth 1999b; Gerber 1996; Hubendick 1950, 1953; Kerney 1999; Kerney and Cameron 1979; Korte and Armbruster 2003; Pilsbry 1948; Sparks 1953.

Vallonia gracilicosta Reinhardt, 1883   Multirib Vallonia

Synonyms: *Vallonia costata montana* Sterki, 1893; *V. albula* Sterki, 1893; *V. sonorana* Pilsbry, 1915.

Description

Shell 2.8 mm wide, flattened-heliciform and translucent pale brownish white or greyish white, with regular, widely-spaced lamellar axial ribs and finer incremental striae between them. Whorls 3 to 3½, the last whorl descending before the aperture. Apertural lip thickened inside by a broad, swollen, opaque-white rib. Genitalia: unknown.

Vallonia gracilicosta is similar to *V. cyclophorella*, which lacks the thick lip rib. Worn shells of *V. gracilicosta* with the axial ribs eroded off may resemble *V. pulchella* and *V. excentrica*, except that the descending final portion of the last whorl.

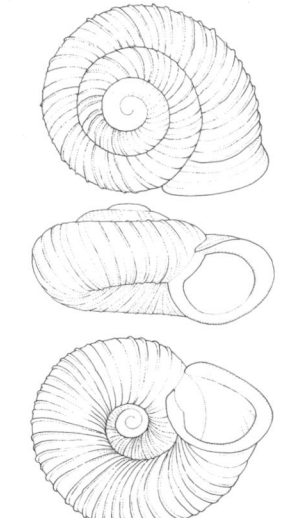

1 mm

Distribution

British Columbia and northern Alberta, east to the Belcher Islands, Nunavut and Newfoundland; south in the U.S. to New York in the east and Arizona, New Mexico and Texas in the west. In B.C., *Vallonia gracilicosta* is recorded from a few scattered localities in the north-central interior and northern mountains. There is an old and questionable record from Vancouver Island.

Natural History

This species lives in leaf litter and under rocks, bark and dead wood, in dry to moist, well-drained places.

Etymology
Vallonia: after the Roman goddess of valleys; *gracilicosta*: "slender ribbed".

Remarks
A recent revision of this genus treated *Vallonia albula* as a subspecies of *V. gracilicosta*, but the differences (such as the relative angle of the apertural lip to the axis of the shell) are minor and not clear-cut.

Selected References
Bequaert and Miller 1973, Dall 1905, Gerber 1996, Pilsbry 1948.

Vallonia pulchella (Müller, 1774)   **Lovely Vallonia**
Synonym: *Helix pulchella* Müller, 1774.

Description
Shell 2.5 mm wide, flattened and translucent, shiny white with fine, irregular incremental striae and occasional low wrinkles. Whorls 3 to 3½: in apical view, they enlarge in a regular spiral to the lip; the last whorl does not descend before the aperture. Apertural lip abruptly flares outward and is broad, flat and thickened within by a rib-like callus. Genitalia: Watson (1920).

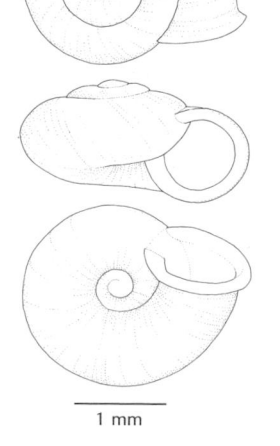

Vallonia pulchella is most similar to *V. excentrica*, but its shell and umbilicus have slightly less elliptical outlines. See the description of *V. excentrica* on page 48.

Distribution
Native to central and eastern North America and central and Europe, north Africa and northern Asia; introduced to many places worldwide, including British Columbia and the western U.S. *Vallonia pulchella* is common in southern B.C. in urban and agricultural areas, but it is less frequently found in the central interior of the province.

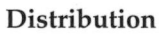

1 mm

Natural History
Vallonia pulchella lives in gardens, parks and roadsides, often with *V. excentrica*, under rocks, concrete, wood, plants and debris. In British Columbia, it is strongly synanthropic. This species appears to prefer wetter habitats than *V. excentrica*, although the two species do sometimes occur together.

Etymology
Vallonia: after the Roman goddess of valleys; *pulchella*: diminutive of "pretty".

Selected References
Bequaert and Miller 1973; Forsyth 1999b; Gerber 1996; Gittenberger et al. 1984; Hanna 1966; Hubendick 1950, 1953; Kerney 1999; Kerney and Cameron 1979; Pilsbry 1948; Sparks 1953.

Family Pupillidae

Pupilla (Pupilla) hebes (Ancey, 1881)   Crestless Column
Synonyms: *Pupa hebes* Ancey, 1881; *Pupilla hebes nefas* Pilsbry & Ferriss, 1910; *P. h. kaibabensis* Pilsbry & Ferriss, 1911; *P. muscorum idahoensis* Henderson & Daniels, 1917; *P. hebes albescens* Ferriss, 1920.

Description
Shell 2.9–3.5 mm high, subcylindrical, with fine, closely spaced, raised wrinkle-like incremental striae. Aperture rounded-ovate, without denticles when viewed straight on (but with a columellar baffle hidden well back behind the columella, and some with a slight bulge in place of a parietal denticle). Apertural lip expanded but not thickened within. Crest low to rather prominent and the same colour as the rest of the shell. Umbilicus narrow. Genitalia: unknown.

Pupilla hebes differs from *P. muscorum* mainly by its lack of a light-coloured crest behind the apertural lip. Compare also with *Lauria cylindracea*, which lacks a crest and has developed denticles within the aperture.

Distribution
Anchorage, Alaska; northwestern British Columbia to northern Mexico and east to Montana and northwestern Wyoming. *Pupilla hebes* is uncommon in B.C., known from a single record in the Cassiar Mountains.

Natural History
Pupilla hebes lives under rocks, wood and vegetation. In the Olympic Mountains of Washington, it has been reported from under Spreading Phlox at and above the timberline.

Etymology
Pupilla: "little pupa", an inferred similarity in shape of the shell to insect pupae; *hebes:* "blunt", a probable reference to the rounded tip of the spire.

Selected References
Beetle 1957, Bequaert and Miller 1973, Branson 1977, Frest and Johannes 2001, Pilsbry 1948, Smith et al. 1990.

Unconfirmed species of Pupillidae

Pupilla (Pupilla) muscorum (Linnaeus, 1758)   Widespread Column

Pupilla muscorum was listed by Aurèle La Rocque (1953) as occurring in British Columbia, but I have not been able to confirm this species existence in the province. It occurs in the Rocky Mountain foothills of Alberta and may be in adjacent parts of B.C. The shell is very similar to *P. hebes,* but the apertural lip is creamy whitish, expanded and thickened within by a prominent ridge, and the crest is light-coloured.

Selected references: La Rocque 1953, Pilsbry 1948.

Family Vertiginidae

Gastrocopta (Albinula) holzingeri (Sterki, 1889)   Lambda Snaggletooth

Synonyms: *Pupa holzingeri* Sterki, 1889; *Bifidaria agna* Pilsbry & Vanatta, 1907.

Description

Shell 1.6–1.9 mm high, subcylindrical and translucent-white. The aperture has six denticles: parietal, columellar, subcolumellar and three palatals. Parietal denticle is forked so that in basal view (with the shell broken away) it forms the shape of the Greek letter lambda (λ), with one end of the fork connected to the apertural lip at the suture. Columellar denticle curves downward; palatals on a callus ridge. Apertural lip thin and expanded, with an oblique crest behind. Genitalia: unknown.

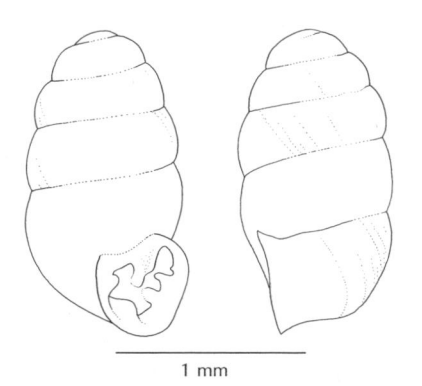

1 mm

The forked parietal denticle is unique among land snails in British Columbia.

Distribution
Ontario and New York west to Montana and Alberta, and south to Kansas and New Mexico. The only records in British Columbia come from Columbia Lake. This species may have a restricted range in B.C., since the genus is almost entirely east of the Rocky Mountains in northwestern North America.

Natural History
In British Columbia, *Gastrocopta holzingeri* has been found in moist leaf litter on a wooded slope.

Etymology
Gastrocopta: "stomach-cut"; *Albinula*: diminutive of "white"; *holzingeri*: after John M. Holzinger, bryologist and botanist from Minnesota.

Selected References
Burch 1966, Forsyth 2004, Hubricht 1985, Oughton 1948, Pilsbry 1948.

Columella columella (von Martens, 1830)   Mellow Column
Synonyms: *Pupa columella* Martens, 1830; *Pupilla alticola* Ingersoll, 1875.

Description
Shell 2.9 mm high, subcylindrical, pale brown, with coarse, irregular incremental striae, with 6 to 7 whorls, the last or second-last whorl often slightly smaller than the adjacent whorls. Aperture lacks denticles and becomes elongated in fully grown shells. Apertural lip thin and not expanded. Genitalia: unknown for North American examples; Pokryszko 1990 for European material.

Columella columella is larger than *C. edentula;* it is much more distinctly cylindrical and adults have an elongated aperture.

Distribution
Arctic and alpine areas of Europe and mainly western North America, south to New Mexico and Arizona, and east to James and Hudson bays, Ontario. In British Columbia, it is recorded from the

Rocky and Cariboo mountains and along the shore of Kootenay Lake. There are records from the Olympic Mountains, Washington.

Natural History
In British Columbia, *Columella columella* occurs in moist montane forests and wet places in valleys, where it lives on vegetation, under dead wood and in litter. In Europe, this species is found in similar habitats, but also in alpine meadows. European populations have both aphallic and euphallic individuals.

Etymology
Columella: "a little column".

Remarks
North American material, when not considered the same as the Eurasian species, is called *Columella alticola*. On geographic grounds, Lothar Forcart and some others treated *C. alticola* as a North American subspecies of *C. columella*, but there are no known morphological distinctions known to separate the species into two subspecies. Other malacologists, including Beata Pokryszko, retained *C. alticola* as a species distinct from *C. columella*.

Selected References
Bequaert and Miller 1973; Forcart 1959a; Kerney 1999; Oughton 1948; Pilsbry 1948; Pokryszko 1987a, 1990.

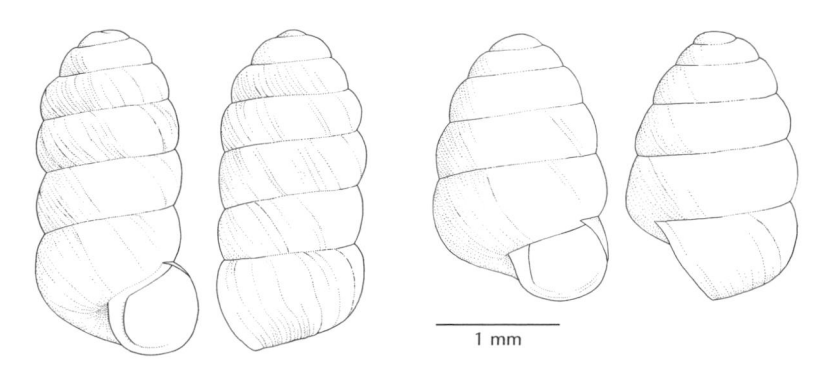

Columella collumella (left) and *C. edentula* (right), described on page 56.

Columella edentula (Draparnaud 1805)   Toothless Column

Description
Shell 2.75 mm high, but usually smaller, subcylindrical but clearly tapering and brown, with coarse and often nearly regular incremental striae. Whorls 5 to 6, the last whorl nearly always larger than the preceding whorl. Aperture lacks denticles and is not prominently elongated when the shell is mature. Apertural lip thin and not expanded. Genitalia: unknown for North American populations; Pokryszko 1990 for European material.

Small juveniles of this species resemble *Punctum randolphii* in size and shape except for a more angular periphery and more rapidly enlarging whorls. Shells of juvenile *Vertigo* species may also be difficult to separate from juvenile *Columella*. Compare this species with *Columella columella*.

Distribution
Northwest Europe, east to the Caucasus and central Asia and much of northern North America. Widespread throughout much of British Columbia, particularly on the coast.

Natural History
In British Columbia, *Columella edentula* lives in all types of forests. During wetter times of the year, it lives predominantly above the ground on the stems and leaves of vegetation, especially Sword Ferns, but it also lives in leaf litter.

Etymology
Columella: "a little column"; *edentula:* "toothless", referring to its lack of apertural denticles.

Remarks
There is some question whether the British Columbia populations are actually *Columella edentula* or another species.

Selected References
Kerney 1999, Kerney and Cameron 1979, Pilsbry 1948, Pokryszko 1987a.

Nearctula species   Threaded Vertigo
Synonym: *Vertigo rowellii* of authors, not Newcombe, 1860.

Description
Shell 2.8 mm high, subcylindrical-ovate, with 5½ to 6 whorls,
dark brown, with close, thread-
like axial riblets and incremental
striae. Aperture has four denticles (parietal, columellar, upper
palatal and lower palatal), but
no palatal callus, sinulus or
crest. Apertural lip thin and
flared. Genitalia: unknown.

This species is recognized by
its large size, coarse sculpture,
simple apertural lip, four denticles, lack of a crest and tapering
spire.

1 mm

Distribution
Southwestern British Columbia to Monterey County, California. In
B.C., it is found in several localities along the east side of Vancouver
Island and near Egmont on the Sunshine Coast.

Natural History
Nearctula species occurs at rich sites in deciduous and mixed forests,
where it lives in moist leaf litter.

Etymology
Nearctula: the Nearctic region.

Remarks
This species has long been called *Vertigo rowelli*, but *Nearctula row-
ellii* now refers to a more southern species formerly known as
Vertigo californica.

Selected References
Cameron 1986, Pilsbry 1948, Rowell 1861, Roth and Sadeghian
2003, Turgeon et al. 1998.

Key to the species of *Vertigo* and *Nearctula*

The species of *Vertigo* are difficult to identify since they are very small and show considerable variation within species. There are at least eight species in British Columbia, but they have received very little study. I include in this key *Nearctula* species (page 57) since it is likely to be confused with species of *Vertigo* and since these two genera are not keyed out separately in the Key to Genera.

1a	Palatal callus typically present, often strong	2
1b	Palatal callus absent or very weak	6
2a	Shell subcylindrical with rib-like incremental striae; columellar denticle sometimes bilobed or buttressed below by a callus	3
2b	Shell subovate, possibly with a longer, tapering spire; fine incremental striae and a columellar denticle not as above	4
3a	Crest huge; columellar denticle sometimes bilobed or buttressed below with a callus	*Vertigo arthuri*
3b	Crest generally less developed; columellar denticle not bilobed	*V. gouldii*
4a	Apertural denticles usually 9	*V. ovata*
4b	Apertural denticles 5	5
5a	Shell subovate	*V.* species
5b	Shell subovate but with the spire longer and distinctly tapered	*V. elatior*
6a	Shell smooth or nearly so	7
6b	Shell rather coarsely regularly lined or with regular riblets (at least on the middle whorls)	8
7a	Shell 2.0–2.5 mm high, usually yellowish brown; apertural lip not flared	*V. columbiana*
7b	Shell more than 2.5 mm high, usually reddish brown; apertural lip flared	*V. andrusiana*
8a	Apertural lip very thin, flared; crest absent	*Nearctula* species
8b	Apertural lip more-or-less thickened, not flared; crest absent to well-developed	9
9a	Axial sculpture coarse, somewhat irregular; angular denticle usually present	*V. modesta*
9b	Axial sculpture relatively regular, almost riblet-like; no angular denticle	*V. cristata*

Vertigo (*Vertigo*) *andrusiana* Pilsbry, 1899   Pacific Vertigo

Description
Shell 2.65 mm high, subcylindrical-ovate, reddish brown and nearly smooth, with fine incremental striae only. Aperture has 4–6 denticles (parietal, columellar, upper palatal and lower palatal always present; angular and subcolumellar sometimes present) and a small palatal callus. Apertural lip straightened or barely indented (no sinulus), thin and flared. Crest absent or low. Genitalia: unknown.

Vertigo andrusiana is similar to *V. columbiana* but reddish rather than yellowish, and slightly larger. *Nearctula* species is larger than *V. andrusiana* and has distinct incremental striae.

Distribution
Vancouver Island, British Columbia to Douglas County, Oregon; San Bernadino Mountains, California.

Natural History
This species occurs in coastal lowland forests.

Etymology
Vertigo: "a whorl"; *andrusiana:* after Fred Andrus, an early Oregon conchologist.

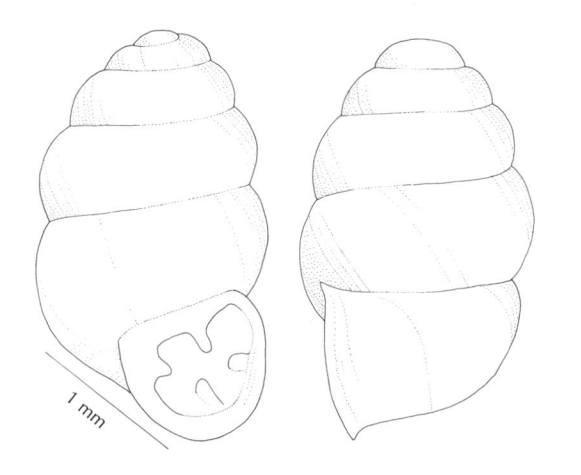

1 mm

Remarks

Malacologists sometimes treat the Californian population as a separate subspecies, *Vertigo andrusiana sanbernardinensis* Pilsbry, 1919; the northern populations would then take the name *V. a. andrusiana*.

Selected References

Cameron 1986, Pilsbry 1948.

Vertigo (Vertigo) arthuri   Callused Vertigo
von Martens, 1882
Synonym: *Vertigo bollesiana arthuri* von Martens, 1882.

Description

Shell 1.75–1.85 mm high, subcylindrical, shiny and orange brown with rather sharp axial riblets and incremental striae. The later whorls are sometimes clearly smaller and more compressed than the middle whorls of the spire. Aperture has six denticles: parietal, angular, columellar, subcolumellar, upper palatal and lower palatal. Angular denticle may be tiny; columellar denticle bilobed or buttressed below by a callus. Palatal callus typically very strongly developed. Apertural lip has a slightly flattened area but otherwise no sinulus. Crest typically very large and swollen. Genitalia: unknown.

This species differs from *Vertigo cristata* and *V. modesta* by the well-developed subcolumellar and often bilobed columellar denticles in addition to a small angular denticle, and by the presence (typically) of a prominent crest and palatal callus. *V. binneyana* Sterki, 1890, has six denticles, like *V. arthuri*, but the shell is smooth

1 mm

and shiny, the crest is never so prominent and the columellar denticle is never bilobed. Compare also with *V. gouldii.*

Distribution
Alaska, Yukon Territory, British Columbia and Alberta east to Ontario and Minnesota and south to South Dakota and Wyoming. Known localities in B.C. are scattered in the Peace River district and in the central interior.

Natural History
In British Columbia, *Vertigo arthuri* has been found under bark in a dry Douglas-fir forest, and in wet litter under Balsam Poplars.

Etymology
Vertigo: "a whorl"; *arthuri:* after Arthur Krause (1851–1920), collector of the original specimens from North Dakota in the 1880s.

Selected References
Burch 1962, Nekola and Massart 2001, Pilsbry 1948.

Vertigo (Vertigo) columbiana   Columbia Vertigo
Pilsbry & Vanatta, 1900

Description
Shell 2.0–2.5 mm high, subcylindrical-ovate, rather glossy and pale yellowish brown; nearly smooth, with only fine, irregular incremental striae that are not developed into well-defined riblets. Aperture has 4–5 denticles (parietal, columellar, upper palatal and lower palatal always present; and sometimes a small angular) and no palatal callus or sinulus. Crest very low. Apertural lip hardly expanded. Genitalia: unknown.

   Vertigo columbiana is smaller than *V. andrusiana* and smoother than most other species in this book.

Distribution
St Paul Island, Aleutian Islands and coast of the Gulf of Alaska south to the Queen Charlotte Islands and Vancouver Island, British Columbia, and Douglas County, southwest Oregon. *Vertigo columbiana* is widespread and common on the west side of the

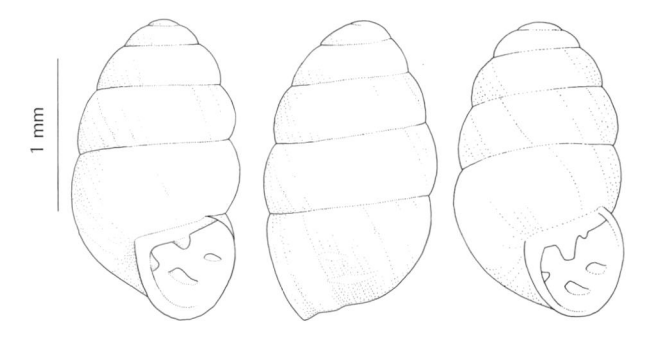

Boundary, Coast and Cascade ranges. North of Vancouver Island, the distribution of this species is poorly known.

Natural History
Vertigo columbiana lives in leaf litter, on ferns and other plants in deciduous and mixed forests.

Etymology
Vertigo: "a whorl"; *columbiana:* "Columbian", referring to British Columbia.

Selected References
Dall 1905, Forsyth 2000d, Pilsbry 1948, Tuthill and Johnson 1969.

Vertigo (Vertigo) cristata Sterki, 1919   Crested Vertigo
Synonyms: *Vertigo gouldii cristata* Sterki, 1919; *V. gouldii* of authors (not A. Binney, 1843).

Description
Shell 2.1–2.2 mm high, subcylindrical and silky, yellowish-brown to orange-brown, marked by fine, more-or-less equal axial striae-riblets. Aperture has four denticles (parietal, columellar, upper palatal and lower palatal), no palatal callus and the sinulus is absent or very weak. Crest low or lacking. Apertural lip thin and scarcely expanded. Genitalia: unknown.

This species differs from *Vertigo modesta* by its smaller size, more regular and sharper axial sculpture, the absence of an angular denticle and often a more nearly cylindrical and more slender form.

Distribution
Probably across much of Canada and parts of the northern U.S. In British Columbia, this species is generally distributed across the northern part of the province in the north-central Interior, and Rocky Mountains.

Natural History
Vertigo cristata inhabits the moist leaf litter in Trembling Aspen and Black Cottonwood groves and in coniferous forests.

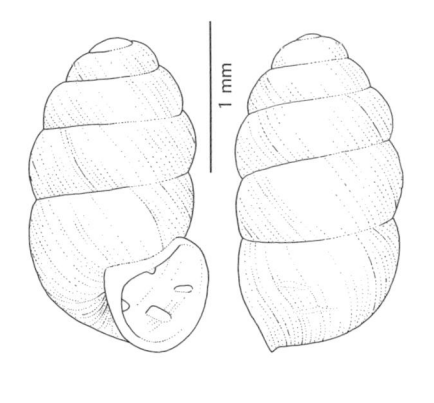

Etymology
Vertigo: "a whorl"; *cristata:* "crested".

Selected References
Nekola 2001, Pilsbry 1948.

Vertigo (Vertigo) elatior Sterki, 1894   Tapered Vertigo
Synonyms: *Vertigo ventricosa elatior* Sterki, 1894; *V. gouldi lagganensis* Pilsbry, 1899; *V. gouldii loessensis* F.C. Baker, 1928.

Description
Shell 2.0–2.2 mm high, subovate, glossy and brown, with a moderately long, distinctly tapered spire; nearly smooth, with only fine incremental striae. Aperture has five denticles: parietal, columellar, subcolumellar, upper palatal and lower palatal. Palatal callus slight or absent; sinulus more-or-less strongly indented. Crest weak or absent. Apertural lip flared and thin. Genitalia: unknown.

This species is most similar to *Vertigo ovata* and *V.* species but the spire is more tapered than either of these and it has fewer denticles than *V. ovata*.

Distribution
Northeastern U.S. and eastern Canada north to Hudson Bay; west to Montana, Alberta and southeastern British Columbia. In B.C., *Vertigo elatior* has only been found at Field, but nearby Lake Louise

(formerly Laggan), Alberta, is the type locality of *V. gouldii lagganensis*, a synonym of *V. elatior*.

Natural History
Nothing is known about the preferred habitat of this species in British Columbia, but is likely a woodland species.

Etymology
Vertigo: "a whorl"; *elatior:* "high" or "lofty".

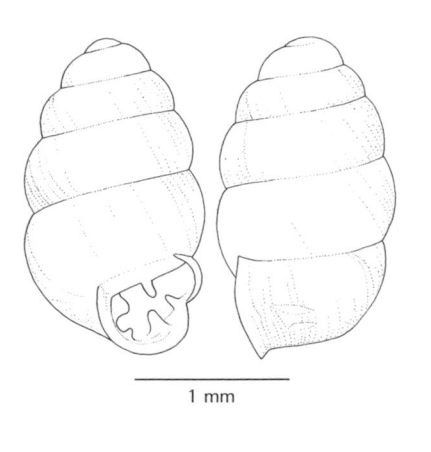

1 mm

Selected References
Hubricht 1984, Metcalf and Smartt 1997, Oughton 1948, Pilsbry 1948, Vanatta 1906.

Vertigo gouldii (A. Binney, 1843)   Variable Vertigo

Description
Shell 1.7–1.9 mm high, subcylindrical, silky yellowish to reddish brown, marked with fine, more-or-less regular axial striae-riblets. Aperture has five denticles (parietal, columellar, upper palatal and lower palatal), a palatal callus and a weak sinulus. Crest moderately developed. Genitalia: unknown.
   Vertigo gouldii resembles specimens of *V. arthuri* in which the crest and palatal callus are not prominently developed; see Remarks.

Distribution
British Columbia and Montana south to New Mexico. In B.C., it occurs in the Rocky Mountains and adjacent ranges.

Natural History
Vertigo gouldii is a common forest-dwelling species.

Etymology
Vertigo: "a whorl"; *gouldii:* after Augustus Addison Gould, a physician in Boston, Massachusetts.

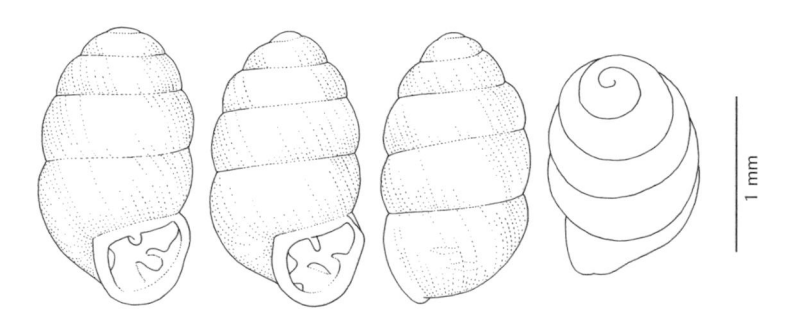

1 mm

Remarks

There is considerable variation in this species, and it could be a species complex. Western North American *Vertigo gouldii* is often subdivided into weakly differentiated subspecies based on the presence or absence of apertural denticles and the strength of the crest or palatal callus. Henry Pilsbry (1948) called specimens from the Rocky Mountains of British Columbia *V. gouldii basidens* Pilsbry & Vanatta, 1900, but apertural denticle characters show great variability within most populations. *V. arthuri* may be an extreme form of this variable species, since they appear to intergrade.

Selected reference: Pilsbry 1948.

Vertigo (Vertigo) modesta (Say, 1824)   Cross Vertigo

Synonyms: *Pupa decora* Gould, 1847; ?*Isthmia corpulenta* Morse, 1865; ?*Pupa corpulenta parietalis* Ancey, 1887.

Description

Shell 2.4–2.9 mm high, subcylindrical-ovate and reddish brown. Aperture has 4–5 denticles (parietal, columellar, lower palatal, upper palatal and usually a small angular), but no palatal callus or sinulus. The curvature of the apertural lip only slightly flattened. Crest low to moderately developed. Apertural lip slightly thickened and not flared. Umbilicus open. Genitalia: unknown.

Vertigo modesta is similar to *V. columbiana* and *V. cristata*, but is slightly larger than both. *V. columbiana* is also more weakly sculptured and *V. cristata* is regularly and strongly sculptured than *V. modesta*.

Distribution
In North America, *Vertigo modesta* occurs from the Aleutian Islands, Alaska, east across northern Canada to Labrador; south to New England and southern California. It is widespread and common, particularly in the northern mountains, including the Rockies.

Natural History
This species lives under stones and dead wood, around and on vegetation, and in leaf litter.

Etymology
Vertigo: "a whorl"; *modesta:* "unassuming".

Remarks
Vertigo modesta, in the broad sense, is often regarded as Holarctic, when the Eurasian *V. arctica* (Wallenberg, 1858) is considered a synonym. There are also several North American forms or possible subspecies in addition to those listed under Synonyms (above) that warrant further study. Most British Columbian examples have an additional small angular denticle and may take another name: *V. modesta parietalis* (Ancey, 1888) or *V. m. corpulenta* (Morse, 1865).

Selected References
Bequaert and Miller 1973, Pilsbry 1948, Pokryszko 1990.

Vertigo (Vertigo) ovata Say, 1822   Ovate Vertigo

Synonyms: *Zonites upsoni* Calkins, 1880; *Pupa ovata antiquorum* Cockerell, 1891; *Vertigo modesta mariposa* Pilsbry, 1919.

Description
Shell 2.5 mm high, subovate, glossy, reddish brown and nearly smooth, with fine incremental striae only. Aperture has nine denticles (angular, parietal, infraparietal, columellar, subcolumellar, suprapalatal, upper palatal, lower palatal and infrapalatal), a heavy palatal callus, a large crest, and a deep and sharply inward-pointing sinulus. Apertural lip thin and flared. Genitalia: described and drawn by Barker (1999) for a *Vertigo* alleged to be *V. ovata*; otherwise unknown.

This species is easily distinguished from all others by its ovate form and its large number of denticles; but see also *Vertigo (Vertigo)* species on the following page.

Distribution
Alaska to Labrador, south to Florida, the West Indies and Baja California, Mexico. It is locally common in suitable habitats throughout British Columbia, both on the coast and in the interior. Gary Barker (1999) reported that *Vertigo ovata* was introduced to New Zealand, but the identification of those records is questionable.

Natural History
This species lives under decaying vegetation, leaf litter and wood in marshes along the shores of waterways. At many localities, *Vertigo*

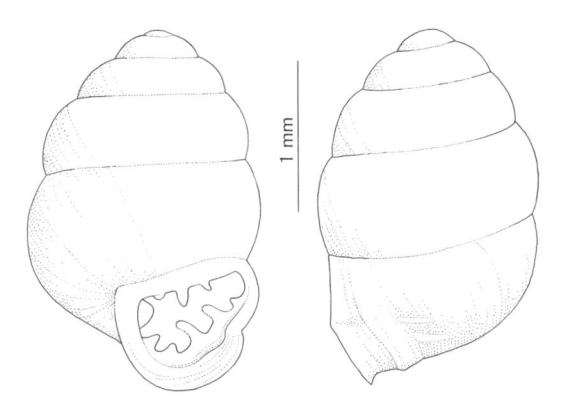

ovata is associated with *Zonitoides nitidus* and *Euconulus praticola*, as well as with succineids. The marshland habitat and wide distribution suggests that migratory waterfowl may play a significant role in the dispersal of this species.

Etymology
Vertigo: "a whorl"; *ovata:* "ovate" or "egg-shaped".

Selected References
Barker 1999, Bequaert and Miller 1973, Pilsbry 1948, Roth and Sadeghian 2003, Smith et al. 1990.

Vertigo (Vertigo) species

Description
Shell 2.2 mm high, subovate, glossy, reddish brown and nearly smooth, with fine incremental striae. Aperture has five denticles (parietal, columellar, subcolumellar, upper palatal and lower palatal), a well-developed palatal callus, a slightly developed crest and a moderately deep sinulus. Apertural lip thin and strongly flared. Genitalia: unknown.

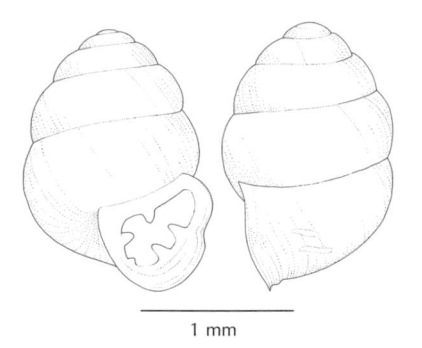

1 mm

This species is closest to *Vertigo ovata*, but has fewer denticles.

Distribution
The only record in British Columbia is from a retail plant nursery near Cobble Hill on southern Vancouver Island, likely introduced.

Natural History
Unknown, since the species is not yet identified. Because it was discovered in a retail plant nursery where it was living in association with a large number of introduced snails and slugs, it may be an exotic species.

Remarks
This may be an unusual form of *Vertigo ovata*, with fewer apertural denticles, or it could be another species, such as *V. ventricosa* (Morse, 1865).

Unconfirmed species of Vertiginidae

Gastrocopta (Vertigopsis) pentodon   Comb Snaggletooth
(Say, 1822)
This species was listed from British Columbia by Aurèle La Rocque (1953), likely based on a record cited by Pilsbry (1898) from "Laggan, B.C.", an error for Laggan (now Lake Louise), Alberta. *Gastrocopta pentodon* lives in southern Canada as far west as Alberta, where it is rare; it is widespread over much of the eastern and central United States. Although known to live on the Alberta side of the Rocky Mountains, I know of no records from B.C. Unlike *G. holzingeri* (see page 53), this species' parietal denticle is not lambda-shaped and the columellar denticle does not curve downward within.
Selected references: Forsyth 2004, Hubricht 1985, La Rocque 1953, Pilsbry 1948.

Vertigo binneyana **Sterki, 1890**   Cylindrical Vertigo
This species was reported from Nanaimo, British Columbia, and Seattle, Washington, but these records were likely based on misidentified material. *Vertigo binneyana* is found in Montana, so it is possible that it could occur in British Columbia, particularly in the southeast. *V. binneyana* is subcylindrical, smooth and shiny; it has six apertural denticles, a palatal callus and a weakly indented sinulus.
Selected references: Burch 1966, Dall 1905, Pilsbry 1948.

Family Haplotrematidae

Ancotrema hybridum (Ancey, 1888)   Oregon Lancetooth
Synonyms: *Selenites vancouverensis hybrida* Ancey, 1888; *S. vancouverensis hybrida* Hemphill, 1890; *Macrocyclias vancouverensis semidecussata* Gratacap, 1901; *Helix sportella* of authors, not Gould, 1846.

Description
Shell 18–27 mm wide, flattened-heliciform, dull or slightly shiny and dark straw-yellow, with a low spire and well-defined axial riblets that are made beaded (especially in and near the umbilicus) by the intersection with fine spiral striae.

Whorls 6, with the final part of last whorl expanded and the axial riblets reduced or absent on the last or second-last whorl. Apertural lip slightly thickened and its upper edge is arched forward and downturned. Animal creamy white with darker head and tentacles; mantle has fine, pale-brown speckles. Colour photograph C-33. Genitalia: not figured; "essentially the same as [*Ancotrema*] *sportella*" (Baker 1930c).

Ancotrema hybridum is intermediate in size between *A. sportella* and *Haplotrema vancouverense*; and *A. sportella* is usually smaller, with beaded sculpture extending on the last whorl to the aperture. *H. vancouverense* has microscopic, closely spaced spiral striae, and while there may be prominent incremental wrinkles, these wrinkles do not resemble ribs and do not intersect with spiral striae to form the beaded sculpture of *Ancotrema* species.

5 mm

Distribution
Aleutian Islands, Alaska, to Humboldt County, California. In British Columbia, *Ancotrema hybridum* is widespread and common on the west side of the Coast Mountains, but H. Burrington Baker listed a record from Vernon, and it should be expected in southeastern B.C., since it occurs in northeastern Washington and northern Idaho.

Natural History
Ancotrema hybridum lives in deciduous, coniferous or mixed-wood forests, scrub, and marshes, where it finds shelter under logs, rocks, vegetation and in humus and leaf litter.

Etymology
Ancotrema: compound of "bend" and "hole", presumably referring to the down-turned lip and aperture; *hybridum*: "hybrid".

Remarks
Ancotrema hybridum was formerly ranked as a subspecies of *A. sportella* until recently when Barry Roth (1990, 1991) gave evidence that it is a distinct species and stated that *A. hybridum* and *A. sportella* occur together without intergradation. Many authors have combined or otherwise confused *A. hybridum* with these two species.

Selected References
Baker 1930c; Frest and Johannes 2001; Roth 1990, 1991.

Ancotrema sportella (Gould, 1846)   Beaded Lancetooth
Synonym: *Helix sportella* Gould, 1846.

Description
Shell 9–22 mm wide, flattened-heliciform and straw-yellow to light greenish, with raised, sharp axial riblets that are cut into bead-like segments by coarse spiral striae running through them; this sculpture continues to the lip in adults. Whorls 6, the final part of the last whorl expanded. Apertural lip slightly thickened and the upper part is arched forward, down-turned by having a crease-like depression. Animal long, slender and creamy white, with darker head and tentacles. The mantle has fine, pale-brown speckles. Genitalia: Pilsbry 1946.

Ancotrema sportella is similar to *Haplotrema vancouverense*, but is smaller, has beaded sculpture and a more down-turned, obliquely creased apertural lip; it also lacks the large, dark blotches that *H. vancouverense* has on its mantle. *A. hybridum* is larger and its axial sculpture becomes obsolete on the last whorl.

Distribution
Aleutian Islands, Alaska, to Humboldt
County, California, east through northern
Washington to northern Idaho. In British
Columbia, widespread and common on the
west side of the Coast Mountains, but
could be expected in the southern interior.

5 mm

Natural History
This species occurs in deciduous, coniferous or mixed forests, in scrub, and (less frequently) in open places. It lives in leaf litter
or in moist places under logs, vegetation
and rocks. It is omnivorous, as are the other
haplotrematids in this book.

Etymology
Sportella: "fruit basket", referring to the
shell sculpture.

Remarks
There is great variability in size of adults between populations,
which suggests that there could be unrecognized species or subspecies present. A subspecies, *Ancotrema sportella sinkyonum* Roth,
1990, is recognized from Humboldt and Mendocino Counties,
California. See Remarks under *A. hybridum.*

Selected References
Frest and Johannes 2001, Porter 1965, Roth 1990.

Haplotrema (*Ancomena*) *vancouverense* Robust Lancetooth
(I. Lea, 1839)
Synonyms: *Helix vancouverensis* I. Lea, 1839; *H. vellicata* Forbes,
1850; *Circinaria vancouverensis chocolata* Dall, 1905.

Description
Shell 22–32 mm wide, flattened-heliciform, dull or slightly shiny,
and yellowish, olive green or darker, with low, unequal incremental
wrinkles and microscopic, closely spaced spiral striae, but no

beaded sculpture. Whorls 5½, the final portion of last whorl expanded. Apertural lip slightly thickened with the upper edge arched forward and straight or slightly down-turned. Animal creamy white, the head and tentacles darker; mantle tan to brown with large dark blotches. Colour photograph C-19. Genitalia: Pilsbry 1946.

Compare *Haplotrema vancouverense* with *Ancotrema hybridum*, which can approach it in size. *H. vancouverense* lacks the beaded sculpture and the crease (or strong impression) near the apertural lip that are present in both species of *Ancotrema*.

5 mm

Distribution
Aleutian Islands and southeast Alaska to northwestern California, and east to northern Idaho and northwestern Montana. In British Columbia it is nearly ubiquitous along the coast and on islands. This species has yet to be recorded from southeastern British Columbia, where it should be expected.

Natural History
Haplotrema vancouverense lives in deciduous, mixed-wood or coniferous forests, and frequently it is common in coniferous forests where other large snail species are absent. It finds shelter under logs, moss and rocks or deep in forest litter. Snails of this family are omnivorous; they eat earthworms, slugs and snails, including their own species.

Etymology
Haplotrema: "simple hole", perhaps referring to the deep, open umbilicus; *Ancomena*: "valley-moon" or "bend-mouth"; *vancouverense*: after Fort Vancouver (now Vancouver), Washington, where it was first collected.

Selected References
Brunson and Osher 1957, Pilsbry 1946, Porter 1965.

Family Testacellidae

Testacella (Testacella) haliotidea   Earshell Slug
Draparnaud, 1801

Description
A large slug, 60–120 mm long, creamy
white, yellow, or dull greyish yellow, with
a small external ear-shaped shell on its posterior end. The shell has a small, spiral protoconch. Two branched grooves extend forward on the body from the anterior edge of
the mantle (beneath the shell). Genitalia:
Quick 1960, Barker 1999.

    This species is unlike any other slug in this book.

Distribution
Native to Europe, from Scotland and Ireland, south to North Africa
and east to the Balkans; introduced to North America, Australia,
New Zealand and elsewhere. In British Columbia, there are several
records in the Royal B.C. Museum of this species from gardens in
Victoria.

Natural History
In British Columbia, *Testacella haliotidea* lives in gardens, but is seldom seen. Its long, slender body with narrowed front end is well
adapted to burrowing. It lives mostly underground and ventures to
the surface only at night. In times of drought, these slugs aestivate
deep underground, but during wet weather, they are more likely to

be found under stones and vegetation at the surface. These slugs are aggressive carnivores that, in the words of Frank Collins Baker, "will pursue an earthworm through its many subterranean burrows … with a persistency that recalls the ferocity of the tiger." These slugs impale their prey – earthworms, molluscs and other invertebrates – with a lance-like radular tooth. They mate underground and lay their eggs there, too. Eggs are relatively large and elongate (7 mm long).

Etymology
Testacella: "little shell"; *haliotidea*: "like a *Haliotis*" (abalone), referring to the open, flat, spiral form of the shell, like that of the abalone.

Selected References
Baker 1902, Barker 1999, Burch 1966, Kerney 1999, Kerney and Cameron 1979, Pilsbry 1946, Quick 1960.

Family Punctidae

Paralaoma servilis (Shuttleworth, 1852)   Pinhead Spot
Synonyms: *Helix servilis* Shuttleworth, 1982; *H. pusilla* Lowe, 1831; *H. caputspinulae* Reeve, 1852; *H. conspecta* Bland, 1865; *Punctum conspectum pasadenae* Pilsbry, 1896; *P. c. alleni* Pilsbry *in* Pilsbry & Ferriss, 1918.

Description
Shell 2.4 mm wide, flattened-heliciform, yellowish brown through olive to dark brown, with a low, rounded spire and 3¼ to 3¾ whorls. Axial ribs more-or-less regularly spaced and sometimes have lamellar edges; fine spiral and incremental striae also present. Apertural lip not thickened. Umbilicus about a quarter the width of the shell. Genitalia: unknown.

Paralaoma servilis most closely resembles *Planogyra clappi*, which has a flatter spire, much higher lamellar axial ribs and a broader umbilicus. *Discus whitneyi* is larger, the axial ribs are not lamellar and the umbilicus is wider. *Punctum randolphii* is much smaller and with densely packed fine axial riblets.

Distribution
Nearly cosmopolitan: parts of Europe, North Africa, Asia, South America, Australia, New Zealand and elsewhere. In North America, from Alaska to Mexico. *Paralaoma servilis* is sporadic and seasonally common throughout British Columbia.

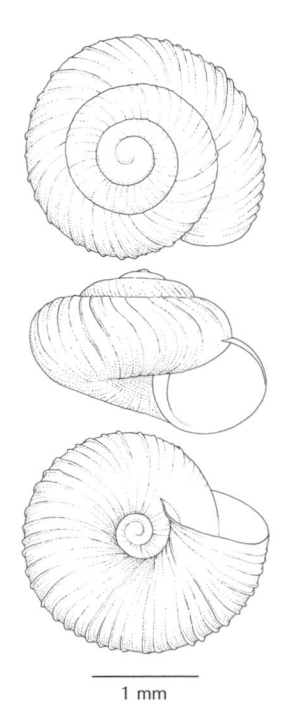

1 mm

Natural History
Paralaoma servilis, while probably native, is mostly found in disturbed sites, under wood or rocks and in leaf litter. It is synanthropic and now recognized as an introduced species in many places worldwide. It is usually seen during cool autumn and spring months.

Etymology
Paralaoma: "akin to *Laoma*", a genus of Pacific-island land snails; *servilis*: "of a slave".

Remarks
In western North America, this species has gone by the names *Punctum conspectum, Toltecia pusilla* and *Paralaoma caputspinulae*. But this species appears to be nearly cosmopolitan and has numerous synonyms and suspected synonyms worldwide. Most recently, Falkner, Ripken and Falkner (2002) pointed out that the correct name should be *Paralaoma servilis*.

Selected References
Bequaert and Miller 1973; Falkner et al. 2002; Gittenberger et al. 1984; Pilsbry 1948; Roth 1985, 1986, 1987b; Seddon and Holyoak 1993; Smith et al. 1990; Smith and Kershaw 1979.

Punctum (*Punctum*) *randolphii* (Dall, 1895)   Conical Spot
Synonym: *Pyramidula randolphii* Dall, 1895.

Description
Shell generally 1.2–1.4 mm wide, but occasionally wider, to 1.8 mm; it is flattened-heliciform, reddish to yellowish brown, with 3¼ to 4¼ whorls, fine raised axial riblets and exceedingly fine spiral striae (most evident in and around the umbilicus). Umbilicus about a fifth to a quarter of the shell width. Body of animal short, hardly extending beyond the shell when crawling. Genitalia: unknown.

Punctum randolphii is smaller than *Paralaoma servilis* and *Planogyra clappi*, and its major axial ribs are not as widely spaced. Several species of *Punctum* thought not to occur in B.C. are very similar – see Remarks.

Distribution
South-central Alaska to the Klamath Mountains, California, and east to Idaho. It is widespread and common along the British Columbia coast, sporadic in the central interior, but common again in the Rockies and other mountain systems of the interior wet belt.

Natural History
Punctum randolphii is gregarious within litter in deciduous, coniferous and mixed-wood forests from sea level to nearly the timberline (1200 m). On the south coast, these snails are ubiquitous in forests and particularly abundant around Bigleaf Maples.

A similar Eurasian species, *Punctum pygmaeum* (Draparnaud, 1805), can reproduce without a mate; although unknown for *P. randolphii*, it is likely that the North American species can also reproduce uniparently.

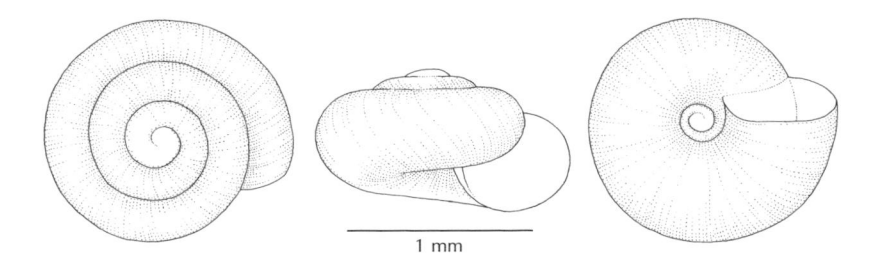

1 mm

Etymology

Punctum: "a dot"; *randolphii*: after P. Brooks Randolph of Seattle, an amateur collector who in the 1890s sent Dall and Pilsbry material from the Puget Sound region and from his travels north to the Klondike during the gold rush of 1897–98.

Remarks

For this book, I regard all British Columbian populations of *Punctum* as *P. randolphii*; but a slightly smaller *Punctum* living in the wet interior mountains and scattered over the central interior may be a different species, *P. minutissimum* (I. Lea, 1841). An eastern and central North American species, *P. minutissimum* is said to occur west to Alberta; it is generally believed to be smaller and often flatter than the west-coast species. A third species, *P. californicum* Pilsbry, 1898, mostly recorded from the southwestern U.S. but with a few records in the northwestern states, is said to be slightly larger. Past attempts to distinguish the North American species from the Eurasian *P. pygmaeum* relied on minor differences in size, proportions of height to width and umbilical width; they did not consider intraspecific variation.

SEM view of *P. randolphii*, showing the densely ribbed surface sculpture.

Selected References

Baur 1987, Forsyth 2001d, Pilsbry 1948, Tuthill and Johnson 1969.

Family Discidae

Anguispira (*Zonodiscus*) *kochi* (Pfeiffer, 1821)

Banded Tigersnail

Synonym: *Helix kochi* Pfeiffer, 1821; *Patula solitaria occidentalis* von Martens, 1882; *Anguispira kochi eyerdami* Clench & Banks, 1939.

Description
Shell 25.5 mm wide, heliciform and brown with a light spiral band just above the periphery; this light band is often bordered by two obscure dark bands. About 6 whorls, with coarse, low axial riblets on early whorls that later become lower and more irregular and less rib-like. Apertural lip not flared and only rarely slightly thickened in older snails. The animal is pale brown or creamy brown with a tinge of ochre, and has darker, greyish brown tentacles. Colour photograph C-1. Genitalia: Pilsbry 1948; Schileyko 2002.

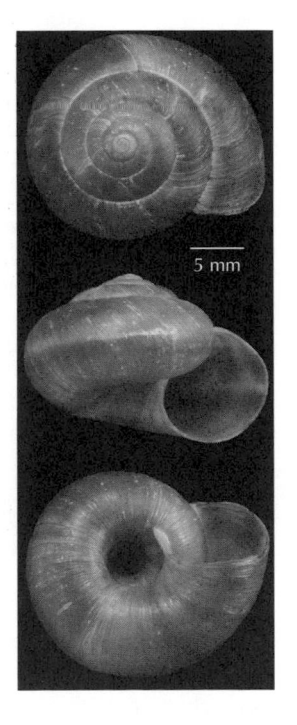

5 mm

The shell of *Anguispira kochi* differs from most *Oreohelix* by being darker, but separation of long-dead and eroded shells may sometimes be difficult. Juveniles can be recognized from juvenile *Allogona ptychophora* – another large snail occurring in the same area – by the larger umbilicus and less angular periphery.

Distribution
Southeast British Columbia and through Washington, Idaho and Montana to Oregon, east of the Cascade Mountains; Lake Erie south and west to Kentucky, Missouri and Arkansas. In B.C. this species is known from the vicinity of the Kootenay River, Kootenay Lake and southern portion of the Columbia River.

Natural History
This species lives in moist, well-vegetated forests, often near the shores of lakes and streams.

Etymology
Anguispira: "snake-spire"; *zonodiscus*: "banded disc"; *kochi*: after someone named Koch.

Remarks
The western population of this species, disjunct from eastern populations, is often treated as distinct subspecies, *Anguispira kochi occidentalis* (von Martens, 1882).

Key to species of *Discus*

1a   Shell brown with more-or-less regularly spaced reddish brown markings .... *Discus rotundatus*
1b   Shell without reddish brown markings 2
2a   Axial ribs continuing onto the basal surfaces of the shell .... *D. whitneyi*
2b   Axial ribs only rarely continuing past the periphery .... *D. shimekii*

Discus (Antediscus) shimekii (Pilsbry, 1890)   Striate Disc
Synonyms: *Zonites shimekii* Pilsbry, 1890; *Pyramidula cockerelli* Pilsbry, 1898; *Zonitoides randolphi* Pilsbry, 1898.

Description
Shell 6.5 mm wide, flattened-heliciform and yellowish brown with fine incremental striae and prominent axial ribs, which rarely extend below the periphery onto the umbilical surface. Whorls 4½, with the periphery of the last whorl rounded. Genitalia: unknown.

    Discus shimekii has a slightly larger and usually paler shell than *D. whitneyi*, whose axial ribs extend onto the umbilical surface.

Distribution
Ogilvie Mountains, Yukon Territory, south into northern British Columbia and along the Rocky Mountains to New Mexico and Arizona; east to central and southeastern Alberta. In B.C. it is locally common in the far north and the Rocky Mountain region.

Natural History
This species is montane, living under rocks, vegetation and dead wood in forests. It is sympatric with the more widespread and abundant *Discus whitneyi*.

Etymology
Discus: "a disc"; *Antediscus*: "prior disc"; *shimekii*: after Bohumil Shimek (1861–1937), who collected the first specimens. Shimek, professor of botany and herbarium curator at the University of Iowa, was a long-time friend of Henry Pilsbry, with whom he began the study of shells as a boy.

Remarks
Somewhat smoother and flatter coiled shells with a wider umbilicus have been named *Discus shimekii cockerelli*, but intermediate forms between *D. shimekii* and *D. shimekii cockerelli* are said to occur. The synonym *Zonitoides randolphii* was named from specimens collected at Lindeman Lake, B.C., during the Klondike gold rush.

1 mm

Selected References
Beetle 1957; Bequaert and Miller 1973; Forsyth 1999a; Pilsbry 1898, 1948.

Discus (Discus) whitneyi (Newcomb, 1864) Forest Disc
Synonym: *Helix striatella* Anthony, 1840; *H. whitneyi* Newcomb, 1864; *H. cronkhitei* Newcomb, 1865; *Pyramidula cronkhitei anthonyi* Pilsbry, 1906.

Description
Shell 6.7 mm wide, flattened-heliciform and light to dark brown. About 4½ whorls, convex or somewhat angular in profile, with prominent axial ribs and fine incremental striae. Apertural lip not thickened or flared. Umbilicus large, about a third the width of the

shell. Animal pale grey on the sides and tail, and dark grey or blackish on the back, head and tentacles. Genitalia: Unknown.

Shells of *Discus whitneyi* vary considerably. The periphery is angled in some, but rounded in others, and the number of axial ribs per whorl varies. Occasional pale coloured shells occur in some populations. In *D. shimekii* the axial ribs fade out at the periphery and seldom extend to the umbilical surface of the last whorl. *Paralaoma servilis* looks like a smaller version of *D. whitneyi*, but its umbilicus is narrower and the initial whorls are spirally lined rather than smooth.

1 mm

Distribution
Aleutian Islands, Alaska, east to Labrador and south to New Mexico and Texas. Common throughout most of British Columbia, but rare on the south coast.

Natural History
Discus whitneyi lives in coniferous and deciduous forests, as well as open sites, and seeks shelter in leaf litter and under logs, vegetation and rocks. It also frequents wet places such as marshes and the edges of waterways.

Etymology
Discus: "a disc"; *whitneyi*: after J.D. Whitney, a member of the California State Geological Survey.

Remarks
This species is better known as *Discus cronkhitei*, which was shown by Barry Roth (1987a) to be a synonym of *D. whitneyi*. But both names may be synonyms of *D. ruderatus* (Férussac, 1821), which occurs throughout Europe and northern Asia.

Selected References
Bequaert and Miller 1973, Hubricht 1985, Pilsbry 1948, Roth 1987a, Roth and Lindberg 1981.

Discus (Patula) rotundatus (Müller, 1774)　Rotund Disc
Synonym: *Helix rotundata* Müller, 1774.

Description
Shell 4.5 mm wide (in British Columbia), flattened-heliciform and yellowish brown with a pattern of more-or-less regularly spaced, reddish spots on its apical surface and with prominent, regular axial ribs and incremental striae. Whorls 5½ to 6, with the last whorl angular at the periphery. Genitalia· J.W. Taylor 1906–14.

The more numerous and compactly coiled whorls and reddish markings readily differentiate this species from *Discus shimekii* and *D. whitneyi*.

Distribution
Western and central Europe, east to Crimea and south to North Africa; introduced to Newfoundland, Ontario, New Jersey, New York, Massachusetts, California, Washington and British Columbia. *Discus rotundatus* is apparently not well established in B. C., with only a single record from Esquimalt on Vancouver Island (1954); there is also a recent record (2001) from Bellingham, Washington.

Natural History
Discus rotundatus lives in varied habitats in Europe, including woodlands, gardens and waste ground. In British Columbia it was found in a garden.

Etymology
Discus: "disc"; *Patula*: "open" or "spreading"; *rotundatus*: "rounded".

Selected References
Kerney 1999, Kerney and Cameron 1979, Pilsbry 1948, Roth 1982b, Seddon and Holyoak 1993.

Family Oreohelicidae

Oreohelix strigosa **(Gould, 1846)   Rocky Mountainsnail**
Synonym: *Helix strigosa* Gould, 1846; *Oreohelix strigosa stantoni* Dall, 1905; *O. strigosa canadica* Berry, 1922. Numerous other infraspecific names have been proposed – see Pilsbry 1939.

Description
Shell 16–26 mm wide, variable in form, but typically flattened-heliciform, opaque and rough textured; pale, greyish white to dark brownish, often with darker bands of brown; variable number of bands, but typi-cally one prominent band above and another just below the periphery, with nar-rower, fainter bands below the lower band on the base (sometimes showing only as faint traces). Spire low-conical to more raised and extremely variable within and between populations. About 6 convex whorls, the periphery weakly to sharply angular. In juveniles the periphery is angu-lar, a character that is retained by some adults or in some populations. Shell sculp-tured with coarse, irregular axial riblets and striae; there may be faint traces of spi-ral sculpture in some. Aperture ovate to rounded; apertural lip barely thickened inside by a low ridge-like callus. Genitalia: Pilsbry 1939.

5 mm

Distribution
Cypress Hills, southeastern Alberta, west to southeastern British Columbia, south through Idaho, Montana, eastern portions of Washington and Oregon to Arizona and New Mexico. In British Columbia, *Oreohelix strigosa* is possibly restricted to the Columbia Basin region and is reported north to Donald Station, the type local-ity of *O. strigosa canadica* Berry, 1922.

Natural History

This species lives near streams, under scattered logs and stones in forests and in vegetated rockslides. B.B. Rees indicated that *Oreohelix subrudis* and *O. strigosa* geographically separated: *O. subrudis* predominates in the streamside habitat, whereas *O. strigosa* occurs among the rockslide rubble. Both Henry Pilsbry and Branley Branson remarked that *O. strigosa* has a markedly discontinuous range in Washington, and that it is restricted mostly to the dry sides of the mountains, where it lives mostly under talus. All species of *Oreohelix* are ovoviviparous.

Etymology

Oreohelix: "mountain snail"; *strigosa*: "full of streaks".

Remarks

Species of *Oreohelix* and populations of snails are differentiated by size, proportions, colour, sculpture and genital anatomy. There is great variation within and between populations, perhaps due to subtle differences in ecology between populations. But the genus has not been adequately studied in British Columbia, treatments of the species in this book are provisional, and there may exist additional species.

Pilsbry distinguished groups of species by their anatomies. In the group containing *Oreohelix strigosa*, the internally folded portion of the penis is less than one-half of its entire length, but in the group of *O. subrudis*, the folded part is greater than one-half the length of the penis.

Selected References

Branson 1980, Pilsbry 1939, Rees 1988, Russell 1951, Solem 1975.

Oreohelix subrudis (Reeve, 1854) Subalpine Mountainsnail

Synonyms: *Helix subrudis* Reeve, 1854; *H. cooperi* W.G. Binney, 1869, in part; *H. limitaris* Dawson, 1875; *Oreohelix cooperi apiarium* Berry, 1919.

Description

Shell 16–23 mm wide, variable in form but typically heliciform or almost beehive-shaped, opaque and rough textured; pale, greyish white to dark brownish, often with a variable number of darker brown bands (as in *Oreohelix strigosa*), and coarse, irregular axial riblets and striae, some with traces of spiral sculpture. Spire moderately raised in typical shells. About 6 convex whorls with a weakly angular periphery; in juveniles the periphery is distinctly angular. Aperture ovate to rounded. Apertural lip barely thickened inside by a low, ridge-like callus. Genitalia: Pilsbry 1939; see Remarks under *O. strigosa*.

5 mm

Distribution
Southeastern British Columbia, south through Washington, Idaho and Montana to Arizona and New Mexico, and east in Canada to the Cypress Hills, which straddles Alberta and Saskatchewan.

Natural History
Oreohelix subrudis lives under small logs, rocks and vegetation in forests and subalpine meadows. It is ovoviviparous.

Etymology
Oreohelix: "mountain snail"; *subrudis*: "somewhat rough".

Remarks
Malacologists recognize several subspecies of *Oreohelix subrudis*. At least some British Columbian populations may belong to the subspecies *O. subrudis limitaris* (Dawson, 1875), which has a dome-shaped shell.

Selected References
Bequaert and Miller 1973, Pilsbry 1939, Russell 1951.

Family Pristilomatidae

Key to species of *Pristiloma*

1a	Umbilicus present	*Pristiloma chersinella*
1b	Umbilicus absent	2
2a	Whorls with distinct axial grooves between the periphery and the suture in addition to very fine axial striae	*P. stearnsii*
2b	Whorls without distinct axial grooves	3
3a	Shell waxy-white or colourless; height of shell less than or equal to half its width	*P. johnsoni*
3b	Shell brownish; height of shell greater than half its width	4
4a	Lip often has an inner denticulate rib; periphery of the last whorl clearly above the middle	*P. lansingi*
4b	Lip always without a denticulate rib within; periphery of the last whorl near median in position	*P. arcticum*

Pristiloma (Priscovitrea) chersinella   Black-foot Tightcoil (Dall, 1866)
Synonym: *Helix chersinella* Dall, 1866.

Description
Shell 3.3 mm wide, flattened-heliciform, glossy and translucent yellowish brown marked only by fine incremental striae. Whorls 4½ to 5 with rounded periphery. Aperture crescent-shaped, with no inner denticulate rib. Umbilicus small. Animal white with black pigmentation on foot and edge of mantle. Genitalia: Pilsbry 1946.

This is the only species of *Pristiloma* in British Columbia whose shell has an umbilicus. The genus *Euconulus* is similar in form but

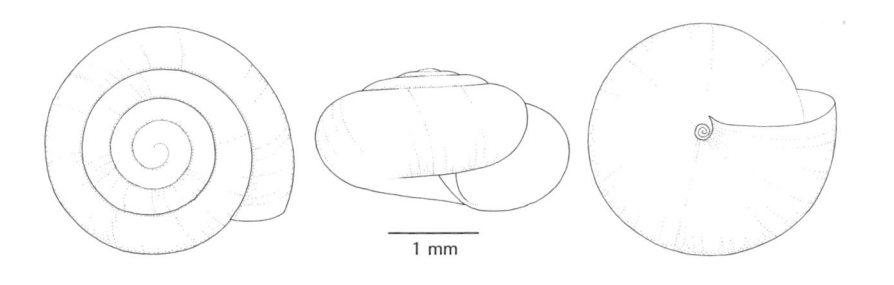

1 mm

brown rather than yellowish and a much smaller, almost closed umbilicus.

Distribution
Mountains of British Columbia, northwestern Montana, northern Idaho, south along the Cascade Range to south-central Oregon, western Nevada and southern California. In B.C., it has been found in the Babine Mountains near Smithers and the Rocky Mountains near and south of Fernie .

Natural History
Pristiloma chersinella is montane, found in British Columbia at 1200–1740 metres above sea level. It lives under rocks in slides and under moss, sticks, bark and logs in wet coniferous subalpine forests and meadows.

Etymology
Pristiloma: "saw-edge", a reference to the saw-tooth lip rib unique to *P. lansingi* (see page 90); *Priscovitrea*: a compound of *prisco*, "primitive" and *vitrea*, meaning "glass", after a genus of Eurasian snails; *chersinella*: diminutive of *chersina*, meaning "dry land".

Selected References
Baker 1931, Berry 1919, Forsyth 2001d, Frest and Johannes 2001, Pilsbry 1946, Roth and Sadeghian 2003.

Pristiloma (Priscovitrea) johnsoni   Broad-whorl Tightcoil (Dall, 1895)
Synonym: *Vitrea johnsoni* Dall, 1895; *Pristiloma taylori* Pilsbry, 1899; *Retinella columna* Morrison, 1937.

Description
Shell 2.8 mm wide, flattened-heliciform, nearly smooth and translucent or waxy-white. Spire nearly flat. About 3½ whorls, rapidly increasing with a rounded periphery. Aperture narrow. No umbilicus. Animal white with black eye spots. Genitalia: Pilsbry 1946.

The small size, colour, rapidly enlarging whorls and lack of an umbilicus are distinctive.

Distribution
Kyuquot Sound, Vancouver Island, British Columbia, to Oregon. Although widespread along the coast, this species is not very common.

Natural History
Pristiloma johnsoni lives in leaf litter of deciduous, coniferous and mixed-wood forests from sea level to over 1300 m in the subalpine, but is seldom encountered in large numbers. It has also been reported from vegetated rockslide habitats.

Etymology
Pristiloma: "saw-edge", a reference to the saw-tooth lip rib unique to *P. lansingi* (see page 90); *johnsoni*: after University of Washington natural-history professor O.B. Johnson.

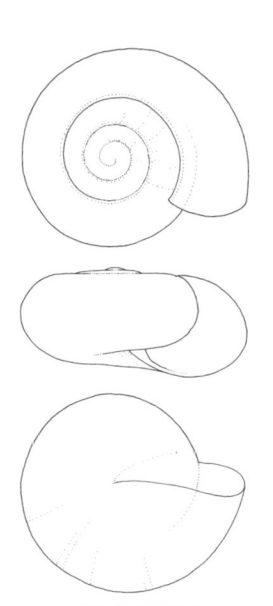

1 mm

Selected References
Baker 1930a, Branson and Branson 1984, Cameron 1986, Pilsbry 1946.

Pristiloma (*Pristiloma*) *lansingi*   Denticulate Tightcoil (Bland, 1875)
Synonym: *Zonites lansingi* Bland, 1875.

Description
Shell 3.0 mm wide, flattened-heliciform, glossy, translucent and pale yellowish brown marked by fine incremental striae and weak grooves. About 5½ whorls, tightly coiled with a slightly angular periphery above the midline of the whorl. Aperture narrow, crescent-shaped and often with a white, irregularly denticulate inner rib. No umbilicus. Animal has a grey head, black tentacles and a whitish foot. Genitalia: Pilsbry 1946.

This species most closely resembles *Pristiloma arcticum*, which has a medial periphery of the last whorl and no denticulate rib-like thickening inside the aperture. *P. lansingi* is the only species of *Pristiloma* with this denticulate thickening.

Distribution
Queen Charlotte Islands, British Columbia, south along the coast to Del Norte County, northern California.

Natural History
Pristiloma lansingi is common and wide-spread, living in leaf litter, under bark and dead wood, in deciduous and mixed forests. It is most common in deep-litter habitats.

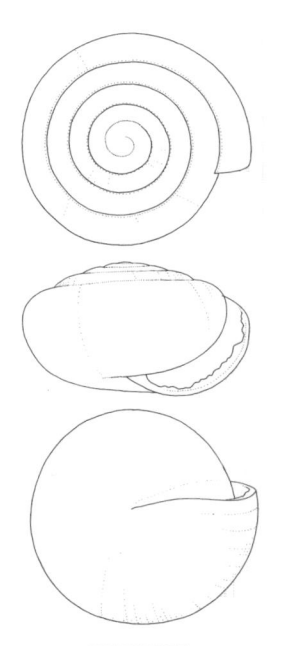

Etymology
Pristiloma: "saw-edge", a reference to the saw-tooth lip rib unique to *P. lansingi*; *lansingi*: after A. Ten Eyck Lansing, a friend of Thomas Bland.

Selected References
Pilsbry 1946, Roth and Sadeghian 2003.

1 mm

Pristiloma (Pristiloma) stearnsii   **Striate Tightcoil**
(Bland, 1875)
Synonym: *Zonites stearnsii* Bland, 1875.

Description
Shell 4.1 mm wide, flattened-heliciform, glossy and translucent dark brown. About 7 tightly coiled whorls with a rounded periphery above the midline. Spire and shoulder of the last whorl marked with axial grooves and ridges that disappear below the periphery. Aperture lacks an inner toothed ridge. No umbilicus. Animal pale grey with darker grey dorsum. Genitalia: Pilsbry 1946.

Juvenile *Pristiloma stearnsii* are often difficult to distinguish from some *P. lansingi*.

Distribution
Southeast Alaska to northwestern Oregon. In British Columbia, *Pristiloma stearnsii* is known only from the west side of the Coast Mountains.

Natural History
This species lives in mixed or coniferous forests in leaf litter or under dead wood and bark. It prefers cold, high-elevation coniferous forests with a shallow litter layer.

Etymology
Pristiloma: "saw-edge", a reference to the saw-tooth lip rib unique to *P. lansingi*; *stearnsii*: after Robert Edward Carter Stearns (1827–1909), a prominent early Californian naturalist.

Selected References
Pilsbry 1946.

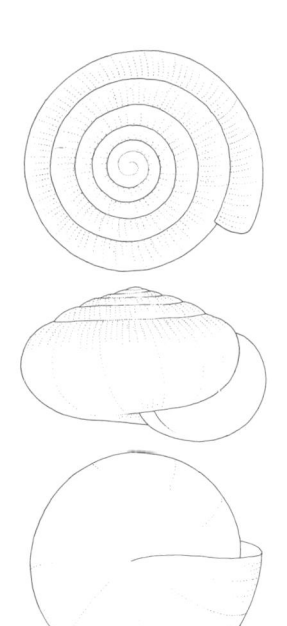

1 mm

Pristiloma (Pristinopsis) arcticum   Northern Tightcoil
(Lehnert, 1884)
Synonym: *Hyalina arctica* Lehnert, 1884; *Pristiloma japonica* Pilsbry & Hirase, 1903.

Description
Shell 3.5 mm wide, flattened-heliciform, glossy and translucent brown and smooth, with fine incremental striae. About 4¾ tightly coiled whorls with a rounded, nearly medial periphery. Aperture crescent-shaped with no inner denticulate rib. No umbilicus. Genitalia: Pilsbry 1946.

    Pristiloma arcticum is similar to *P. lansingi*, whose periphery is above the middle of the whorl; and in *P. lansingi* a denticulate rib is often evident inside the apertural lip.

Distribution
Aleutian Islands and the Arctic Slope of Alaska, south to Idaho, Washington and southern Oregon; Kamchatka, Russia; Hokkaido and the Kurile Islands, Japan. Currently this species is known in British Columbia from a few collections made in the Skeena and Cascade mountains.

Natural History
Pristiloma arcticum is montane and lives under rocks and vegetation in wet sub-alpine forests, meadows, seeps and bogs.

Etymology
Pristiloma: "saw-edge", a reference to the saw-tooth lip rib unique to *P. lansingi* (see page 90); *pristinopsis*: "primitive-appearance"; *arcticum*: "northern" or "arctic".

Remarks
There is a rather weakly differentiated sub-species – *Pristiloma arcticum crateris* Pilsbry, 1946 – known from Crater Lake, Klamath County, Oregon.

Selected References
Forsyth 2001d, Frest and Johannes 2001, Pilsbry 1946, Roth and Lindberg 1981.

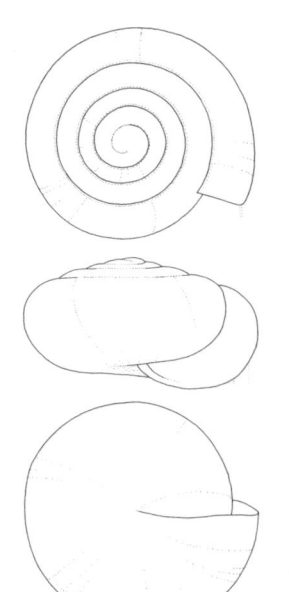

1 mm

Vitrea (Crystallus) contracta (Westerlund, 1871)　　　Contracted Glass-snail

Synonym: *Zonites crystallina contracta* Westerlund, 1871.

Description
Shell 2.6 mm wide, flattened-heliciform, glossy, translucent and colourless, nearly smooth, with only fine incremental striae. Almost 4 tightly coiled whorls with a low spire. Umbilicus about a sixth the width of the shell. Animal pale tan to ivory white, with transparent ocular tentacles and black eye spots. Genitalia: Lohmander 1938.

Vitrea contracta appears similar to *Microphysula cookei* and *Hawaiia minuscula* – all have colourless, rather tightly coiled shells – but both of them have a smaller umbilicus and no spiral striae.

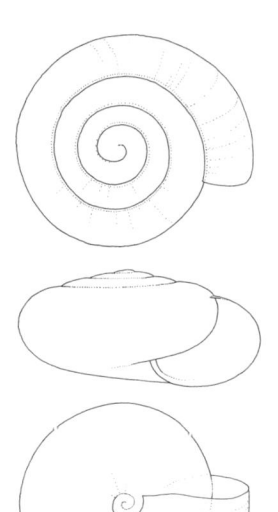

1 mm

Distribution
Central and northwest Europe; introduced to the Middle East, Australia and North America, where it is known from California, Washington, British Columbia and Ontario. In B.C., it is recorded on Vancouver Island, from Nanaimo to Victoria, on the Gulf Islands, and in the lower Fraser Valley and Greater Vancouver.

Natural History
Vitrea contracta is synanthropic and occurs in association with other such species near human settlements. In British Columbia, it lives in gardens and open sites under rocks, wood and debris, and in leaf litter and on other dead vegetation. In Europe, it occurs in calcareous grasslands and woodlands, taluses, walls and caves.

Etymology
Vitrea: "glassy"; *Crystallus:* "a crystal"; *contracta:* "contracted", referring to the narrow aperture.

Selected References
Forsyth 1999b, Kerney 1999, Kerney and Cameron 1979, Kuiper 1964, Roth 1977, Roth and Pearce 1984, Smith and Kershaw 1979, Valovirta and Väisänen 1986.

Unconfirmed species of Pristilomatidae

Hawaiia minuscula (A. Binney, 1841)　　　Minute Gem
In 1905, William Healy Dall listed this species from Victoria and Departure Bay on Vancouver Island, but these old records are unconfirmed and likely erroneous. I do not consider this species to be native to British Columbia. The nearest records in the U.S. are from southeast of Sequim, Washington and from Central Point in southwest Oregon. Branley Branson's record of *H. miniscula* from Sequim is based on a misidentification. Terrence Frest and Edward Johannes record it as rare and possibly introduced in Idaho.
Selected references: Branson 1977, Dall 1905, Frest and Johannes 2001, Hanna 1966, Pilsbry 1946.

Family Euconulidae

Euconulus (*Euconulus*) *fulvus* (Müller, 1774)　　　Brown Hive
Synonyms: *Helix fulva* Müller, 1774; *Conulus fulvus alaskensis* Pilsbry, 1899.

Description
Shell 3.5 mm wide (slightly wider than its height), heliciform, translucent brown with a silky lustre, marked only by fine, close axial threads and exceedingly fine spiral striae. On the base, the axial threads are fainter and the spiral striae often more apparent (but sometimes absent). About 5½ whorls, with a clearly conical spire and a rounded or slightly angular periphery. Aperture narrowly crescent-shaped. Umbilicus absent or tiny. Animal greyish with darker tentacles; mantle has large black blotches visible through the shell. The foot long and slender. Genitalia: Pilsbry 1946.
　　Compare this species with *Euconulus praticola*.

Distribution
Holarctic: across Europe, Asia, North Africa and all of North America, south to Sinaloa, Mexico. Introduced to Australia. Ubiquitous throughout British Columbia.

Natural History

Euconulus fulvus is common and wide-spread, living under logs and debris, and in dead grass and leaf litter at both dry and moist sites at all elevations, both inland and along the coast. Reproduction in this genus of snails is probably largely by self-fertilization.

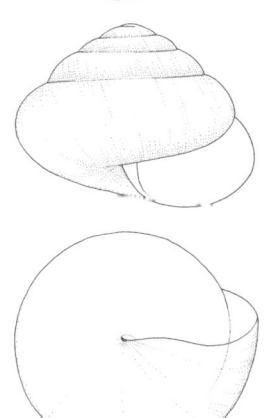

Remarks

A western North American subspecies, *Euconulus fulvus alaskensis* (Pilsbry, 1899), is sometimes recognized. It is said to differ from *E. fulvus fulvus* by having one less whorl in shells of similar size and minor sculptural discrepancies, but many authors have disputed these differences. The validity of *E. fulvus alaskensis* is unresolved and the name is treated as a synonym.

Etymology

Euconulus: "good little cone"; *fulvus*: "tawny" or "reddish yellow".

1 mm

Selected References

Bequaert and Miller 1973, Falkner et al. 2002, Kerney 1999, Pilsbry 1946, Stanisic 1981, von Proschwitz 1985.

Euconulus (*Euconulus*) *praticola* (Reinhardt, 1883)   Marsh Hive

Synonyms: *Hyalina (Conulus) praticola* Reinhardt 1883; *Helix fulva alderi* of authors, not J.E. Gray *in* Turton, 1840; *Euconulus fulva* of authors, not Müller, 1774.

Description

Shell 3.5 mm wide and similar to *Euconulus fulvus*, but darker and reddish brown, with the apical surfaces of the whorls shiny rather than silky. Under about 50× magnification or more, the spiral striae and axial grooves on the basal side of the shell are more distinct than those of *E. fulvus*. Animal almost black. Genitalia: unknown.

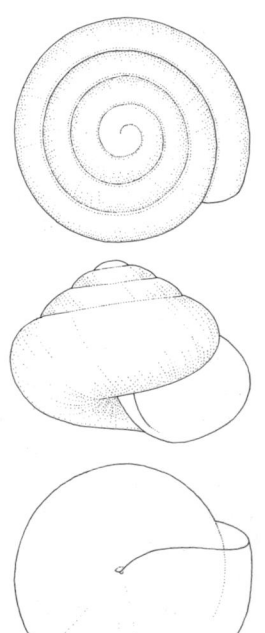

Distribution

Euconulus praticola ranges from Scandinavia, the British Isles and Spain to central Europe, but it has been confused with other species of *Euconulus* and its distribution is likely more extensive. In central and southern British Columbia, *E. praticola* is sporadically distributed near the coast and in the interior. Records of *E. alderi* from Ontario and the midwestern U.S. are probably this species.

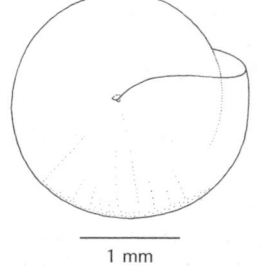

Natural History

This species is characteristically a calciphile, occurring in wet habitats, including marshes, shores of waterways and saturated meadows, where it lives among grasses and other vegetation and on wood.

1 mm

Etymology

Euconulus: "good little cone"; *praticola*: "dwelling in meadows".

Remarks

I use the name *Euconulus praticola* for a species in British Columbia that is clearly separable from *E. fulvus* by the characters of shell, animal and habitat. The name *Euconulus alderi* had been applied to the wetland dwelling *Euconulus* in Europe, North Africa and parts of the U.S. Midwest and Ontario; but recently, Falkner, Ripken and Falkner treated *E. alderi* as a synonym of another forest-dwelling species and resurrected the name *E. praticola* for the wetland species in Europe. Nevertheless, A.A. Schileyko and some other recent authors believe that there is probably only one variable species: *E. fulvus*.

Selected References

Falkner et al. 2002, Kerney 1999, Kerney and Cameron 1979, Schileyko 2002, von Proschwitz 1985.

Family Gastrodontidae

Striatura (Pseudohyalina) pugetensis (Dall, 1895)

Northwest Striate

Synonym: *Patulastra pugetensis* Dall, 1895; *Radiodiscus hubrichti* Branson, 1975.

Description
Shell 1.8 mm wide, flattened-heliciform and translucent, pale yellowish or yellowish brown. Whorls 2½ to 3; spire low. Protoconch has fine, close spiral threads. Subsequent whorls have close, regular, oblique axial riblets and much finer spiral and axial striae (see the SEM images on the next page). Umbilicus is more than a third the width of the shell. Animal translucent white, with a small black blotch over the lung and grey head and tentacles. Genitalia: Pilsbry 1946.

The pale yellowish shell, large umbilicus and intricate sculpture readily distinguish this species.

Distribution
Queen Charlotte Islands, British Columbia, to California and Isla Guadalupe, Baja California, Mexico, and in western Montana. East of the Coast Mountains in British Columbia, this species is known from the North Thompson River Valley (Cariboo Mountains). It was also recorded on the Hawaiian island of Kauai.

Natural History
Striatura pugetensis lives in moist leaf litter of deciduous, coniferous and mixed forests, from sea level to elevations of nearly 700 metres.

Etymology
Striatura: referring to the grooved or striate shell sculpture; *Pseudohyalina*: "false glass"; *pugetensis*: after the Puget Sound region.

1 mm

SEM images of *Striatura pugetensis* showing the intricate microsculpture. The image on the left shows the axial striae.

Remarks
Radiodiscus hubrichtii Branson, 1975, was described from the Olympic Peninsula, but soon after was recognized as a synonym of *Striatura pugetensis*.

Selected References
Baker 1941; Branson 1975; Berry 1919; Forsyth 2001d; Pilsbry 1946; Smith et al. 1990; Solem 1977a, b.

Zonitoides (Zonitoides) arboreus (Say, 1816)   Quick Gloss
Synonym: *Helix arborea* Say, 1816; *H. breweri* Newcomb, 1864.

Description
Shell 5.6 mm wide, flattened-heliciform, dark brown and shiny, with irregular, weak incremental wrinkles and microscopic spiral striae (visible with reflective light and magnification of at least 50×). Whorls 4 to 4½. Animal bluish grey on the tentacles and dorsum, lighter on the sides and tail. Genitalia: Barker 1999, Pilsbry 1946.

Zonitoides arboreus is similar to *Z. nitidus*, but lacks the orange spot on the mantle and has a slightly flatter spire. It is also similar to *Aegopinella nitidula*, which is larger, has coarser spiral striae and a larger umbilicus. Species of *Oxychilus* are smoother and glossier with no spiral striae.

Distribution
North America, south to Central America and the West Indies; introduced into South America, Europe, islands in the North Atlantic,

northwest and southern Africa, the Middle East, Japan, Kamchatka, Australia and Hawaii. It is nearly ubiquitous throughout British Columbia, but is most common east of the Coast Mountains.

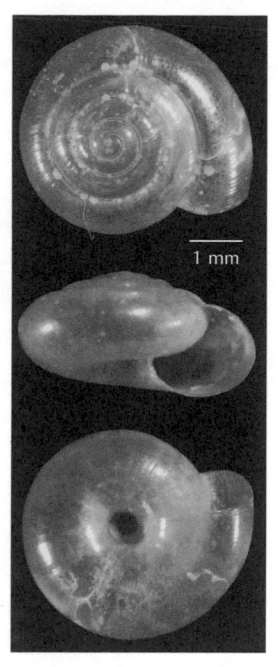

1 mm

Natural History
In British Columbia, *Zonitoides arboreus* lives in a wide variety of dry to wet, mostly natural habitats, under rocks and in rotten wood, leaf litter and vegetation. In many countries where this snail is introduced, it is a serious horticultural and agricultural pest.

Etymology
Zonitoides: "resembling *Zonites*", a genus of snails in Europe; *arboreus*: "of the trees".

Selected References
Barker 1999, Bequaert and Miller 1973, Karlin 1956, Kerney and Cameron 1979, Pilsbry 1946.

Zonitoides (*Zonitoides*) *nitidus* (Müller, 1774)   Black Gloss
Synonym: *Helix nitida* Müller, 1774.

Description
Shell 5.9 mm wide, flattened-heliciform, dark brown and shiny, with irregular, low wrinkle-like axial striae. Whorls 4¾ to 5. Animal black, with a dull orange spot on the mantle. When the animal is retracted into its shell, this spot can be seen through the shell behind the apertural lip, between the suture and the periphery. Genitalia: Pilsbry 1946.

See the comparisons in the descriptions of similar species, *Zonitoides arboreus* and *Aegopinella nitidula* (page 101).

Distribution
Native to Europe, North Africa, northern Asia, Iceland and most of North America; introduced to Australia and Madeira. In British

Columbia, this species is locally common in suitable habitats; although recorded from the lower Fraser Valley, Vancouver Island and the Okanagan Valley, it is not known from central or northern B.C. Henry Pilsbry proposed that *Zonitoides nitidus* might be an exotic along the west coast; he may be correct, since its distribution in B.C. suggests that it is introduced.

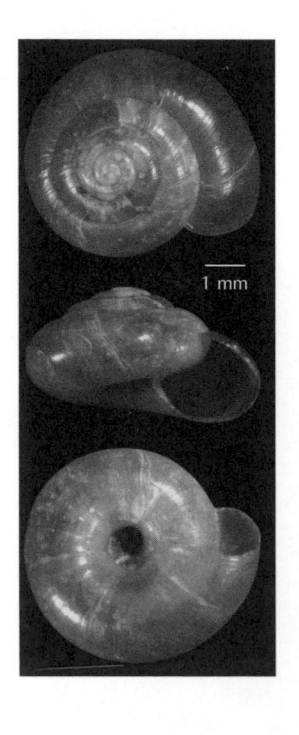

1 mm

Natural History
Zonitoides nitidus is a calciphile, characteristic of marshes and wet areas along the edges of rivers, sloughs, lakes and ponds, where it lives under wood, rocks and vegetation. These snails are omnivores; they eat vegetation and gastropod eggs, including those of their own species. Reproduction is thought to be mostly by self-fertilization, and the species is well known for the variability of the genitalia, which may be either euphallic or hemiphallic.

Etymology
Zonitoides: "resembling *Zonites*", a genus of snails in Europe; *nitidus*: "shining".

Selected References
Adam 1960, Jordaens et al. 1998, Kerney 1999, Kerney and Cameron 1979, Moens 1990, Pilsbry 1946, Platt 1980, Seddon and Holyoak 1993, Watson 1934.

Family Daubebariidae

Aegopinella nitidula **(Draparnaud, 1805)   Waxy Glass-snail**
Synonym: *Helix nitidula* Draparnaud, 1805.

Description

Shell 8–10 mm wide, flattened-heliciform
and yellowish brown, but more opaque
and whitish around the umbilicus; it has a
waxy lustre, rather than glossy, with microscopic, close spiral striae and incremental
striae. Animal dark grey with a slight
bluish tinge dorsally, paler flanks and a
pale sole. Genitalia: Forcart 1959b, Gittenberger et al. 1984.

5 mm

Aegopinella nitidula is similar to *Oxychilus* species, but has exceedingly fine spiral
microsculpture and a broader umbilicus. In
Zonitoides arboreus, the spiral microsculpture is even shallower and closer than that
of *A. nitidula.*

Distribution

Native to Europe; introduced to North
America where it has been found only in
Vancouver and Burnaby, British Columbia.
Earlier records of this species from
California and the Northwest Territories
are now considered erroneous.

Natural History

In Greater Vancouver, this synanthropic species occurs in gardens
and garden-waste dumps where it lives under leaf litter, dead
wood, logs, stones and debris. In Europe, it lives in woods and
grasslands and in association with humans. The life cycle of
Aegopinella nitidula is mainly biennial, maturing and breeding in the
summer following its second winter. Its diet is principally dead
plant material, but it also eats small snails, slugs, earthworms and
other invertebrates.

Etymology
Aegopinella: a diminutive of *Aegopis* (a genus of European snails), which means "goat meadow"; *nitidus:* "shining".

Selected References
Cameron 1982; Forsyth et al. 2001; Hermida et al. 1995; Kerney 1999; Kerney and Cameron 1979; Mordan 1977, 1978; Riedel 1957, 1983.

Nesovitrea (*Perpolita*) *binneyana* (Morse, 1864)   Blue Glass
Synonym: *Retinella binneyana occidentalis* H.B. Baker, 1930.

Description
Shell 3.7 mm wide, flattened-heliciform, translucent, glossy and almost colourless or with a greenish or greenish yellow tinge; it has narrow, widely spaced axial grooves and microscopic, closely spaced spiral striae (though the striae are difficult to see in some shells, even at 50× magnification). About 3½ rapidly enlarging whorls. Genitalia: Pilsbry 1946.

Similar to *Nesovitrea electrina*, which is larger, darker and does not have the extremely faint spiral striae.

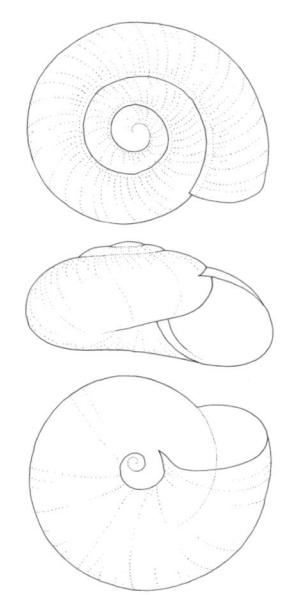

Distribution
British Columbia to Colorado, east to Maine and Quebec; widespread throughout B.C.

Natural History
Nesovitrea binneyana lives in forests, under rocks and dead wood, and in leaf litter.

1 mm

Etymology
Nesovitrea: a compound of *neso*, meaning "island" (a reference to the Hawaiian Islands from where the type species of the genus occurs), and *vitrea*, meaning "glass", after a genus of similar appearing

Palaearctic snails; *Perpolita*: "polished all over"; *binneyana*: after Amos Binney (1803–47), physician, malacologist and founder of the Boston Society of Natural History.

Remarks
The shell of the supposed western subspecies, *Nesovitrea binneyana occidentalis*, is said to differ from the subspecies *N. b. binneyana* by having stronger spiral striae, which are nevertheless very fine and faint. The status of this subspecies remains unresolved.

Selected References
Baker 1930b, Hubricht 1985, La Rocque 1953, Pilsbry 1946.

Nesovitrea (*Perpolita*) *electrina* (Gould, 1841) Amber Glass
Synonym: *Helix electrina* Gould, 1841; *Hyalina pellucida* Lehnert, 1884; *Nesovitrea hammonis* of authors, not Strøm, 1765.

Description
Shell 4.9 mm wide, flattened-heliciform, translucent and glossy yellowish to brownish, marked with narrow, widely spaced axial grooves, but no fine spiral striae. Whorls 3¼ to 3¾, rapidly increasing in width. Animal darkly pigmented; head, back and tentacles almost black; sole of the foot dark grey. Genitalia: Pilsbry 1946.

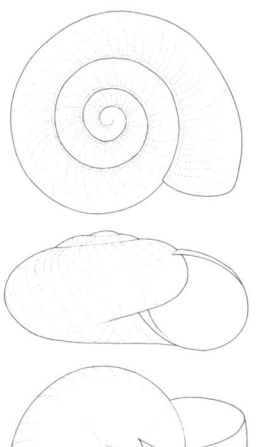

Similar to *Nesovitrea binneyana*, which is smaller, paler in colour and has microscopic spiral striae.

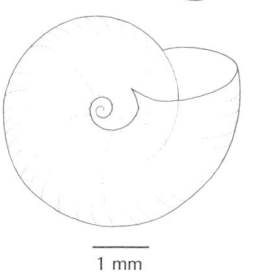

Distribution
Kodiak Island, Alaska, to Labrador and south to Arizona. Throughout British Columbia, but less common than *Nesovitrea binneyana*.

Natural History
Nesovitrea electrina lives in forests where it finds shelter under rocks and dead wood

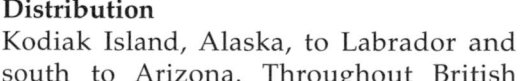

1 mm

and in leaf litter. In the eastern U.S., inhabits wetter places than *N. binneyana*.

Etymology
Nesovitrea: a compound of *neso*, meaning "island" (a reference to the Hawaiian Islands from where the type species of the genus occurs), and *vitrea*, meaning "glass", after a genus of similar appearing Palaearctic snails; *Perpolita*: "polished all over"; *electrina*: "amber", describing the translucency and colour of the shell.

Selected References
Bequaert and Miller 1973, Hubricht 1984, Pilsbry 1946.

Key to species of *Oxychilus*

1a   Emits a strong garlic odour when disturbed; shell less than 7 mm wide...*Oxychilus alliarius*

1b   Never emits a garlic odour; shell not less than 7 mm wide... 2

2a   Shell 10 mm wide; last whorl expands evenly; animal pale grey with minute brown specks on the edge of the mantle (evident near the pneumostome on the right side, and on the left side through the shell) ....*O. cellarius*

2b   Shell 12–16 mm wide; last whorl expands rapidly near the aperture so that it appears more than twice as wide as the previous whorl; animal dark bluish grey............*O. draparnaudi*

Oxychilus (*Oxychilus*) *alliarius*   Garlic Glass-snail
(J.S. Miller, 1822)
Synonym: *Helix alliaria* J.S. Miller, 1822.

Description
Shell 6–7 mm wide, flattened-heliciform, glossy and translucent brown, but slightly more opaque around the umbilicus; nearly smooth, with some fine incremental striae; 5 to 6 whorls. Apertural lip thin and unexpanded. Umbilicus about a sixth to a quarter the width of the shell. Animal dark grey to almost black. It emits a strong garlic odour when handled. Colour photograph C-24.

Genitalia: Barker 1999, Gittenberger et al. 1984, Giusti and Manganelli 2002.

Oxychilus alliarius is most easily distinguished from other species by the garlic odour and pigmentation of the animal. *Zonitoides arboreus* and *Z. nitidus* are more coarsely striated, less glossy and have more elevated spires. *Aegopinella nitidula* is larger and has fine spiral striae and a proportionally wider umbilicus.

Distribution
Western Europe, but introduced to North America, Greenland, Australia, New Zealand South Africa and elsewhere. *Oxychilus alliarius* is widespread and common in populated areas of southwestern British Columbia. In the north, it has also been found at a disturbed site in the Queen Charlotte Islands.

Natural History
Oxychilus alliarius is synanthropic and gregarious, living in moist gardens, parks and modified habitats where it finds shelter under wood, rocks, vegetation and debris. In Europe it also occurs in forests. The strong garlic odour is believed to be a defensive response that the snail produces when disturbed. The odour is present throughout the year and even in very young snails. It is emitted from a thick, brown mucus secreted from an area close to the pneumostome on the mantle. *O. alliarius* eats vegetation primarily, but also preys on other snails and their eggs.

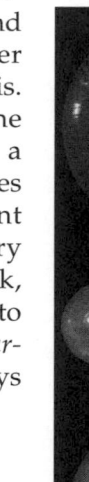

1 mm

Etymology
Oxychilus: "sharp-lip"; *alliarius*: "garlic".

Selected References
Barker 1999; Forsyth 1999b; Kerney 1999; Kerney and Cameron 1979; Lloyd 1970a, 1970b, 1970c; Pilsbry 1946; Robinson 1999; Roth and Pearce 1984.

Oxychilus (Oxychilus) cellarius (Müller, 1774)

Cellar Glass-snail

Synonym: *Helix cellaria* Müller, 1774.

Description

Shell 7–10 mm wide, flattened-heliciform, nearly smooth, glossy and translucent pale yellow-brown, paler and more opaque around the umbilicus; 5 to 6 whorls. Umbilicus about a sixth the width of the shell. Animal grey on the head, tentacles, back and tip of the tail, with paler sides. Mantle suffused with tiny brown speckles around the pneumostome. A row of exceedingly fine brown specks runs along each side just above a groove parallel to the edge of the foot. Genitalia: Barker 1999, Gittenberger et al. 1984, Giusti and Manganelli 1997.

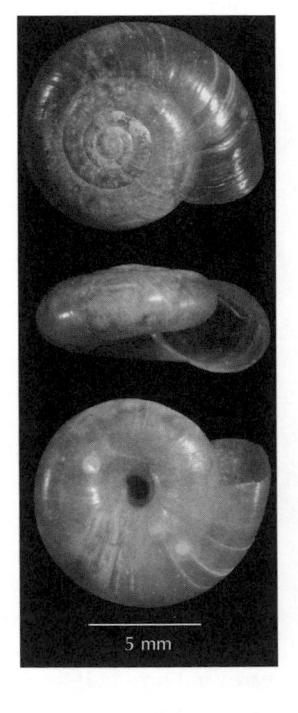

5 mm

The *Oxychilus cellarius* shell is larger than that of *O. alliarius* and smaller than that of *O. draparnaudi*, and the spire is generally flatter than the *O. alliarius* spire. Its body is paler than both the other species, and it does not emit a garlic odour.

Distribution

Western and central Europe and the western Mediterranean region. Introduced to North America, South America, Australia, New Zealand, South Africa and the Philippines. In British Columbia, it is found only in the suburban areas of southern Vancouver Island and the lower Fraser Valley.

Natural History

This synanthrope is common in parks, gardens and modified habitats, where it lives under rocks, wood and vegetation. *Oxychilus cellarius* has a roughly annual life cycle with most breeding activity in autumn. It lays oval eggs, about 1.5 mm in diameter, white and calcareous. *O. cellarius* is omnivorous, and not as carnivorous as *O. draparnaudi*. It is known to feed on the eggs of *Arion* species, plant material, slugs, snails, eggs and earthworms.

Etymology

Oxychilus: "sharp-lip"; *cellarius*: "of the cellar" – the species was first described from shells found in the wine cellars of Copenhagen.

Selected References

Barker 1999, Cameron 1982, Forsyth 1999b, Kerney and Cameron 1979, Lloyd 1970c, Pilsbry 1946, von Proschwitz 1994, Rigby 1963.

Oxychilus (*Oxychilus*)   Dark-bodied Glass-snail
draparnaudi (Beck, 1837)

Synonym: *Helix lucidus* Draparnaud, 1801; *H. draparnaudi* Beck, 1837.

Description

Shell 16 mm wide, flattened-heliciform, pale brown, nearly smooth, but with irregular incremental wrinkles and striae; 5 to 6 whorls. Umbilicus about a sixth the width of the shell. Animal long, slender and bluish black or dark bluish grey, but paler on the sides. Colour photograph C-20. Genitalia: Barker 1999, Gittenberger et al. 1984, Giusti and Manganelli 1997.

    Oxychilus draparnaudi is the largest of the three *Oxychilus* species in British Columbia. The shell is less glossy and translucent than other *Oxychilus* species, and around the umbilicus it is slightly more flattened and more opaque than in the other two species. The animal's pigmentation is distinctive and it has no garlic odour.

5 mm

Distribution

Mediterranean region and western Europe. Introduced to Russia, parts of Africa, North America, Australia, New Zealand and Asia. In British Columbia, it is recorded from southern and eastern Vancouver Island, the Gulf Islands, Greater Vancouver and the lower Fraser Valley. In the western United States, it has been found in Washington, Oregon, California, Colorado and Arizona.

Natural History
Oxychilus draparnaudi is a common synanthropic species in parks, gardens and modified habitats, where it lives under rocks, wood, vegetation and debris. It feeds on the eggs of slugs, small snails, worms, green plants, feces, carrion and fungi.

Etymology
Oxychilus: "sharp-lip"; *draparnaudi*: named for the French zoologist and botanist Jacques-Philippe-Raymond Draparnaud (1772–1804), who wrote on the non-marine molluscs of France.

Selected References
Forsyth 1999c, Frest and Rhodes 1982, Kerney 1999, Kerney and Cameron 1979, Lloyd 1970c, Pilsbry 1946, Proschwitz 1994, Robinson 1999.

Family Milacidae

Unconfirmed species

Milax gagates **(Draparnaud, 1801)**   **Greenhouse Slug**
Native to the Mediterranean region and western Europe, *Milax gagates* is a scattered and rare introduction in the U.S. Pacific Northwest and Idaho, and it may be present in British Columbia. **Selected references:** Carl and Guiguet 1972, Frest and Johannes 2001, Pilsbry 1948.

Family Vitrinidae

Vitrina pellucida (Müller, 1774) Western Glass-snail

Synonyms: *Helix pellucida* Müller, 1774; *Vitrina pfeifferi* Newcomb, 1861; *V. alaskana* Dall, 1905.

Description

Shell 6.2 mm wide (usually less), heliciform, very fragile, translucent, glossy, and greenish, yellowish green or nearly colourless. Spire small and low. Whorls 2½ to 3, rapidly enlarging so that the last whorl forms much of the shell. Apertural lip is thin and delicate. Umbilicus tiny. Animal brownish, especially near the tentacles and on the head; a small lobe on the right side of the mantle partially overlaps the shell when the animal extends itself. Genitalia: Bequaert and Miller 1973, Forcart 1955, Roth and Lindberg 1981.

No other species in this book is likely to be confused with *Vitrina pellucida*. It differs from the related *V. angelicae* Beck, 1837 (Eastern Glass-snail), by having its penis enclosed in a sheath (see figure 5, page 7).

Distribution

Western North America, south to southern California, Arizona and New Mexico; across Europe and northern Asia. *Vitrina pellucida* occurs throughout British Columbia but is more widespread and common away from the coast and in the north. The eastern limit of its range in Canada is unknown, because empty shells of this species are indistinguishable from those of its eastern relative, *V. angelicae*. Some or all records of *V. angelicae* from Alberta may actually be of *V. pellucida*.

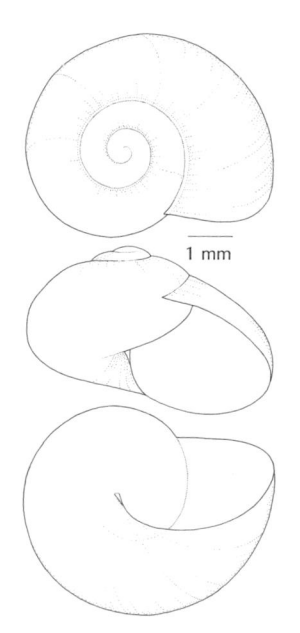

1 mm

Natural History

Vitrina pellucida occurs in alpine tundra, rockslides, roadsides and grassy fields. These snails often occur in seasonally dry places from sea level to high elevations but are never found in forests. They find shelter under rocks and wood, and in grass and leaf litter. This is a cold-weather, annual species usually seen alive in the early spring as juveniles, sometimes when remnants of snow and ice remain, and then in the fall as adults. But empty shells are found more often than live animals. *V. pellucida* is carnivorous.

Etymology
The genus and species names both refer to the glassy, translucent shell.

Remarks
This species is best known in western North America under the synonymous name *Vitrina alaskana* Dall, 1905. Lothar Forcart's comparisons of the reproductive anatomy in *Vitrina* species, published in 1955, became the foundation for later studies that synonymized *V. alaskana* with *V. pellucida*.

Selected References
Bequaert and Miller 1973, Forcart 1955, Kerney 1999, Kerney and Cameron 1979, Pilsbry 1946, Roth and Lindberg 1981, Roth and Sadeghian 2003.

Family Boettgerillidae

Boettgerilla pallens **Simroth, 1912**   **Wormslug**
Synonym: *Boettgerilla vermiformis* Wiktor, 1959.

Description
This slender, worm-shaped slug is 30–60 mm long and 3–5 mm wide when active. It is pale translucent grey, bluish grey or yellowish grey, with the keel, back and head darker bluish grey in adults; juveniles are nearly white. Mantle a third the length of the body, pointed behind, with a pattern of fine concentric wrinkles and a groove approximately parallel to the right edge; the pneumostome is near the midpoint. Keel extends from the tip of the tail to the mantle. Colour photograph C-12. Genitalia: Gittenberger et al. 1984.

This distinctive slug is unlikely to be confused with any other species in this book.

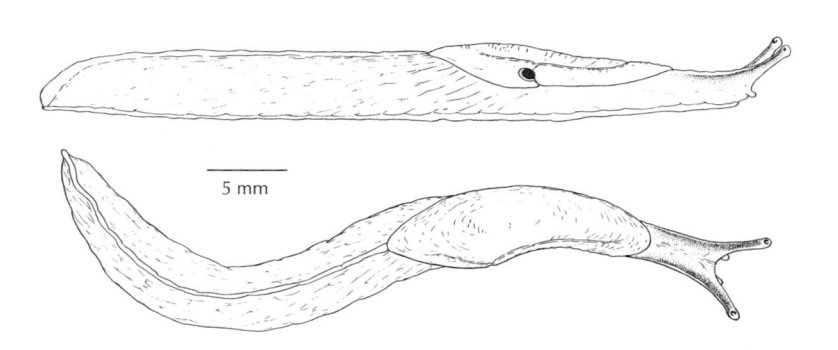

Distribution
Native to the Caucasus Mountains of southeastern Europe. It has spread into western Europe, from Scandinavia south to the Pyrenees Mountains. Introduced to British Columbia, *Boettgerilla pallens* has been found on the Saanich Peninsula and in Oak Bay, Cobble Hill and Metchosin, all on Vancouver Island. It is also introduced to Colombia, South America.

Natural History
This unusual worm-like slug is well adapted for moving through soil and the burrows of earthworms. It has been found living to depths of 60 cm. Although predominantly subterranean and averse to daylight, a small portion of each population lives at the surface, under wood, leaf litter and stones. In British Columbia, this species is recorded from semi-wild urban parks and from a retail plant nursery. It likely also lives in city gardens, but its burrowing habit and relatively small size help it go unnoticed. In Europe, *Boettgerilla pallens* mates and lays eggs underground during the late summer and autumn. It feeds on earthworm feces, detritus, soil surfaces, decaying vegetation, roots, fungal hyphae, carrion and eggs of other slugs.

Etymology
Boettgerilla: after Oscar Boettger (1844–1910), German malacologist who worked and published on Caucasian slugs; *pallens:* "pale".

Selected References
Borreda et al. 1996, Colville et al. 1974, Gunn 1992, Gittenberger et al. 1984, Hausdorf 2002, Kerney 1999, Kerney and Cameron 1979, Reise et al. 2000, Van Goethem 1972.

Family Limacidae

Limacus flavus **(Linnaeus, 1758)**   **Yellow Gardenslug**
Synonyms: *Limax flavus* Linnaeus, 1758.

Description
This large slug is 75–100 mm long, yellowish with grey mottling. Mantle also mottled grey, with a fine fingerprint-like pattern of ridges; pneumostome behind the midline. Tentacles pale bluish. Keel near the end of the tail only. Body mucus yellowish; foot mucus colourless. Sole of the foot is yellowish white. Third intestinal loop short; caecum long. Colour photograph C-32. Genitalia: bursa copulatrix duct opens into the oviduct; illustrated in Barker 1999 and Quick 1960.

Limacus flavus is similar to the larger *Limax maximus*, but the coloration differs. *L. maximus* is grey with blackish spots, never yellow, and usually has distinct lateral bands and reddish brown tentacles. These species can be further differentiated anatomically by their genitalia and intestines.

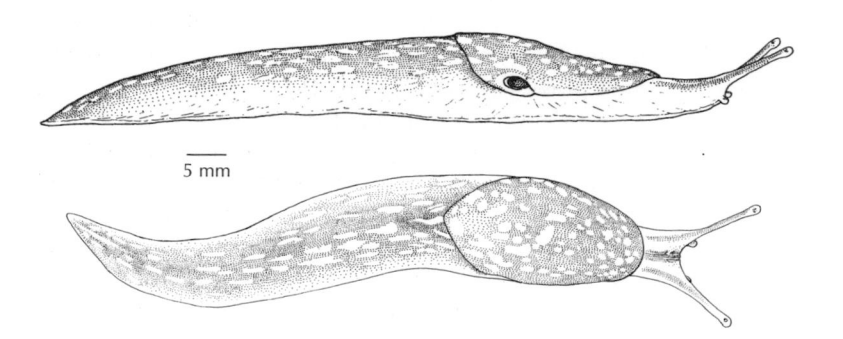

5 mm

Distribution
Southern and western Europe; North Africa east to the Middle East. Introduced to many places, including North America, southern Africa, Australia, New Zealand, China and Japan. In British Columbia, it has been found in North Vancouver and on the Saanich Peninsula, Vancouver Island. In the western U.S., *Limacus flavus* is reported from Washington, California and Arizona.

Natural History

In British Columbia, *Limacus flavus* is found in gardens and compost bins. In Europe, this species lives in woods but is more common in gardens, cellars and crevices of stone walls and buildings where it seeks out moist, dark places. It eats decaying plant material, mildew, fungi and lichens.

Etymology

Limacus: "slug"; *flavus:* "yellow".

Selected References

Barker 1999, Bequaert and Miller 1973, Branson 1977, Chelazzi et al. 1988, Hanna 1966, Kerney 1999, Kerney and Cameron 1979, Pilsbry 1948, Quick 1960.

Limax maximus Linnaeus, 1858   Giant Gardenslug

Description

This large slug can grow to a length of 100 mm or more. It is yellowish grey or brown and spotted or striped with black, but in many the spots coalesce into two or three pairs of lateral bands, often interrupted. Mantle irregularly spotted or mottled with brown (never banded) and has a fine fingerprint-like pattern of ridges. Pneumostome in the right posterior margin of the mantle. Tentacles reddish brown. Keel near the tip of the tail only. Sole creamy white. Mucus colourless. Third intestinal loop long; no intestinal caecum. Colour photograph C-30. Genitalia: penis long and cylindrical, without an appendix; illustrated in Barker 1999, Gittenberger et al. 1984 and Quick 1960.

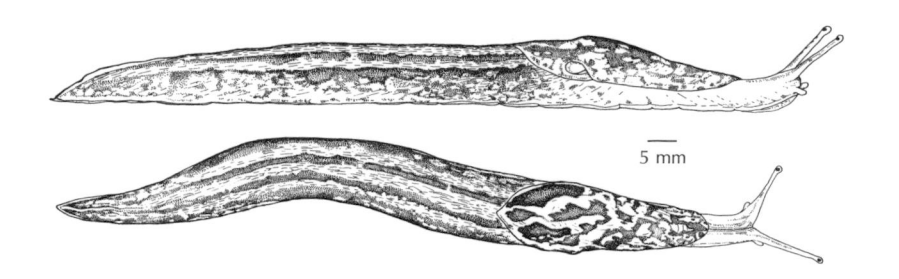

5 mm

Limax maximus is similar to *Lehmannia valentiana*, which is smaller, lacks the spots on the mantle and normally has a stronger banding pattern. See also the comparison with *Limacus flavus* (page 112).

Distribution
Native to Europe, Asia Minor and North Africa. Introduced to North America, South America, Hawaii, Australia and New Zealand. In British Columbia, this slug has been found in the Okanagan Valley, Kootenays, Cariboo and Fraser Valley, in Greater Vancouver, on the Gulf Islands, and in Victoria and elsewhere on southern Vancouver Island. It is also introduced to southeastern Alaska.

Natural History
Limax maximus lives in gardens and wooded areas near human settlement where it seeks out damp, shaded places, under rocks, wood and vegetation. It is nocturnal, venturing out from shelter during the night; a well-developed homing behaviour enables it to return to its home site. *L. maximus* eats mainly fungi and decaying plant matter; green plants do not form a major part of its diet. These slugs mate while suspended from an overhead surface on a thread of mucus. Eggs are oval, about 5.5 mm in diameter and laid in clusters in the spring and autumn. *L. maximus* can live for three to four years.

Etymology
Limax: "slug"; *maximus:* "largest".

Selected References
Barker 1999, Bequaert and Miller 1973, Hanna 1966, Kerney 1999, Kerney and Cameron 1979, Pilsbry 1948, Rollo and Wellington 1975, Roth and Pearce 1984.

Lehmannia valentiana   Three-band Gardenslug
(Férussac *in* Férussac & Deshayes, 1822)
Synonyms: *Limax valentiana* Férussac *in* Férussac & Deshayes, 1822; *L. poirieri* Mabille, 1883; *L. marginatus* of authors, not Müller, 1774.

Description
This slug can grow up to about 70 mm long. It is pale yellowish grey or yellowish violet, somewhat translucent and thin-skinned, and usually has one dark band on each side of the midline and some-times a second weaker pair of bands below. Keel short and not extending forward to the mantle. Mantle has a fine fingerprint-like pattern of ridges. Pneumostome in the posterior right margin of the mantle. Sole is pale grey. Mucus watery, colourless and not sticky. Third intestinal loop short; intestinal caecum long. Colour photo-graph C-31. Genitalia: penis has a blunt-ended finger-like flagellum; see also Barker 1999, Gittenberger et al. 1984 and Quick 1960.

Similar to species of *Deroceras*, *Limax* and *Limacus*, but they all lack bands on the mantle, are less translucent and differ anatomi-cally. The bands on *Lehmannia valentiana* vary among individuals and may be very pale on some, particularly large adults.

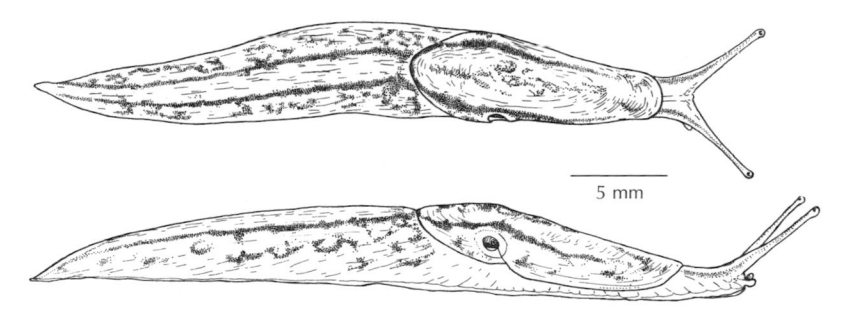

5 mm

Distribution
Native to the Iberian Peninsula, but now widely introduced throughout Europe and worldwide, including North America, South America, several Oceanic islands, Australia, New Zealand and South Africa. In British Columbia, *Lehmannia valentiana* is not widespread, perhaps because of its intolerance to cold; it has been recorded only from southern Vancouver Island and the Gulf Islands.

Natural History
Lehmannia valentiana is nocturnal and prefers moist habitats. It feeds on decayed wood and green plants. As a defence against predation, it can secrete a great amount of mucus. Eggs are translucent yellow, oval and 2.25 mm in diameter.

Etymology
Lehmannia: after the German malacologist Rudolph Lehmann (1812–71); *valentiana*: for the type locality at Valencia, Spain.

Selected References
Barker 1999, Forsyth 2001c, Hanna 1966, Hommay et al. 1998, Kerney 1999, Kerney and Cameron 1979, Waldén 1962, Webb 1961.

Family Agriolimacidae

Key to species of *Deroceras*

1a   Body mucus milky white when the animal is irritated; body colour cream or pale grey, usually with darker grey reticulation; animal 35–50 mm long; penial flagellum lobed and branched, but not with five or six long, slender, nearly equal branches ....*Deroceras reticulatum*

1b   Body mucus always clear; body colour varies; animal up to 35 mm long; penial flagellum absent or consisting of five or six long, slender, nearly equal branches.....................................2

2a   Penis has a 5–6-branched flagellum; animal up to 35 mm long and brown, often with darker spots and fine speckles ....*D. panormitanum*

2b   Penis without a flagellum; animal up to 25 mm long, variable in colour and possibly spotted...3

3a   Bursa copulatrix duct enlarged and has longitudinal folds inside ....*D. hesperium*

3b   Bursa copulatrix duct not enlarged and without internal folds..*D. laeve*

Deroceras (*Deroceras*) *hesperium*   Evening Fieldslug
Pilsbry, 1944
Synonyms: *Deroceras hyperboreus* of authors, not Westerlund, 1876; ?*Limax laevis* Müller, 1774.

Description
A small slug, about 20 mm long, light brown with scattered light and dark spots. No intestinal caecum. Genitalia: penis long and cylindrical, swollen at the anterior end and lacking a medial constriction, its terminal gland simple and sac-like with a small wart at its apex; no penial flagellum; stimulator (within the penis) small and conical; inner wall of penis proximal to the stimulator has two unequal longitudinal folds; bursa copulatrix duct much enlarged, with a rounded or irregularly shaped cavity lined with longitudinal folds. See Pilsbry 1948.

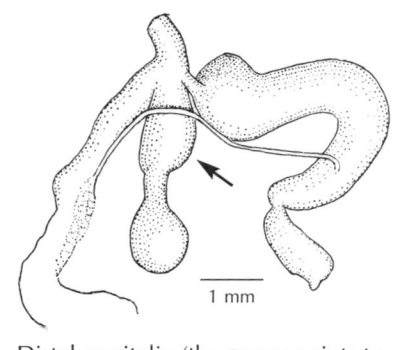

    The key feature to look for is the enlarged bursa copulatrix duct (indicated by the arrow in the illustration) with multiple longitudinal folds inside.

Distal genitalia (the arrow points to the bursa copulatrix duct). There is no image of the whole animal.

Distribution
Comox, Vancouver Island, south to the Columbia River, Oregon; west of the Cascade Mountains in Washington and Oregon.

Natural History
Deroceras hesperium lives in low-elevation, mixed-wood forests in Washington. Otherwise, we know little about this species' natural history; we do not know if it is aphallic, like many individuals of *D. laeve*.

Remarks
Andrzej Wiktor recently questioned the status of *Deroceras hesperium*, suggesting that it may be a synonym of *D. laeve*. Gary Barker had earlier placed *D. hesperium* in the synonymy of *D. laeve*. *D. hesperium* was described as a new species on characters of the

distal genitalia of slugs from Oregon and Vancouver Island. The only record of this species in British Columbia was collected in 1887, and no other slugs recognizable as *D. hesperium* have been found since in Canada. Some supposed records of this species in adjacent regions (e.g., Washington records by Branley Branson) need confirmation.

Etymology
Deroceras: "neck-horn"; *hesperium:* "of the evening", a reference to the western distribution of this species.

Selected References
Barker 1999, Branson 1977, Pilsbry 1948, Wiktor 2000.

Deroceras (Deroceras) laeve (Müller, 1774)   Meadow Slug
Synonyms: *Limax laevis* Müller, 1774; *L. campestris* A. Binney, 1842; *L. campestris occidentalis* Cooper, 1872; *L. montanus* Ingersoll, 1875; *L. castaneus* Ingersoll, 1875; *L. hyperboreus* Westerlund, 1876; *Agriolimax montanus typicus* Cockerell, 1888; *A. montanus intermedius* Cockerell, 1888; *A. montanus tristis* Cockerell, 1888; *Limax hemphilli* W.G. Binney, 1890; *L. hemphilli pictus* W.G. Binney, 1892; *Agriolimax campestris zonatipes* Cockerell, 1892; *A. hemphilli ashmuni* Pilsbry & Vanatta, 1910; ?*Deroceras hesperium* Pilsbry, 1944; ?*D. monentolophus* Pilsbry, 1944.

Description
A small slug, about 25 mm long, dark brown to nearly black. Mucus colourless. No intestinal caecum. Colour photograph C-21. Genitalia: usually aphallic or hemiphallic, though sometimes euphallic; when present, the penis is long and has no medial contraction, it has wart-like structures at its proximal end, its stimulator is hemispherical to bluntly conical, the inner wall has two unequal ridges, the bursa copulatrix duct is narrow and not lined with longitudinal folds, and the vas deferens enters the penis two thirds of the way along its length. Also illustrated in Barker 1999, Gittenberger et al. 1984, Kerney and Cameron 1979, Pilsbry 1948, Quick 1960 and Wiktor 2000.

Compare with similar species, *Deroceras hesperium* and *D. panormitanum.*

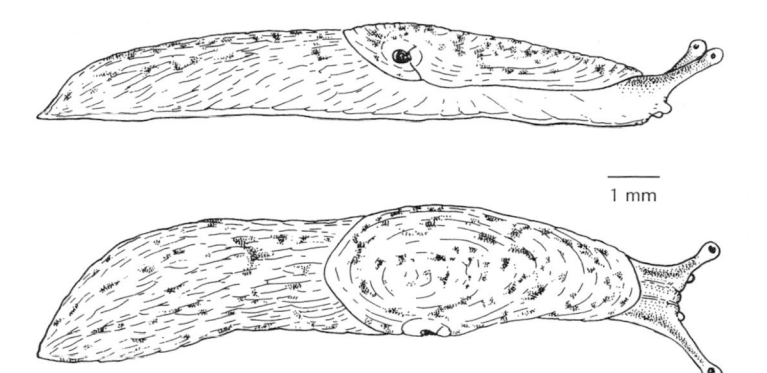

Distribution
Native to Europe, Asia and the Americas from the Arctic to the high Andes; introduced to many places worldwide. Although obviously native throughout British Columbia, some populations, such as those in greenhouses, may be exotic.

Natural History
Deroceras laeve is usually found in wet meadows, marshes and woods, but also in greenhouses. It is a relatively fast-moving species. N.L. Nicklas and R.J. Hoffman claimed that this species is parthenogenic, but Richard Lebovitz argued in favour of self-fertilization. While there is good evidence for outcrossing, other evidence suggests predominant or exclusive self-fertilization, at least in some populations. Reproduction occurs year round.

Etymology
Deroceras: "neck-horn"; *laeve*: "smooth".

Selected References
Adam 1960, Bequaert and Miller 1973, Cameron et al. 1983, Foltz et al. 1982, Getz 1959, Hanna 1956, Hoffmann 1983, Lebovitz 1998, Likharev and Rammel'meier 1962, Nicklas and Hoffman 1981, Pillard 1985, Rollo and Wellington 1975, Shen 1995.

Distal genitalia.

Deroceras (*Deroceras*) *panormitanum* Longneck Fieldslug
(Lessona & Pollonera, 1882)

Synonyms: *Limax panormitanum* Lessona & Pollonera, 1882; *Agriolimax caruanae* Pollonera, 1891.

Description

This slug is 25–30 mm long, chocolate-brown or grey-brown, finely speckled or flecked with darker brown; thin-skinned and somewhat translucent. Mantle paler over the lung. Sole pale grey. Mucus is thin, colourless and not especially sticky. No intestinal caecum. Genitalia: penis two-lobed with 4–6 long, slender, unbranched penial appendixes originating between the lobes; stimulator (within the penis) bluntly conical. See also Barker 1999, Gittenberger et al. 1984, Kerney and Cameron 1979, Pilsbry 1948, Quick 1960 or Wiktor 2000.

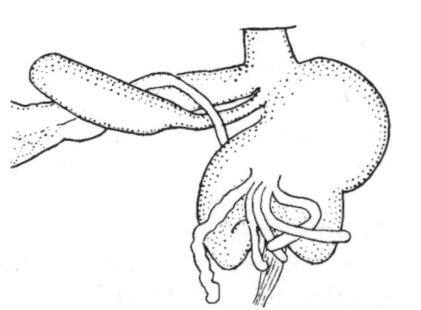

This species is likely to be confused with *Deroceras laeve* (and *D. hesperium*), but it is larger. It is best identified by the structure of the penis and its accessory organs.

Distal genitalia.

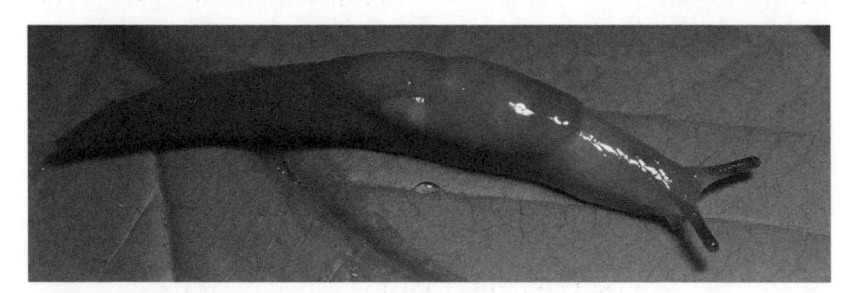

See the colour version of this photograph, C-22.

Distribution

Native to southwestern Europe; introduced to many parts of Europe, North America, South America, Australia, New Zealand and South Africa. In British Columbia, it is found in Greater Vancouver and Victoria.

Natural History

Deroceras panormitanum is synanthropic and lives in gardens and greenhouses under vegetation and other suitable cover. These slugs eat green and decaying plant material. They are fast moving and noted for their cannibalistic and aggressive behaviour of tail lashing and biting. *D. panormitanum* mates year round and produces oval eggs about 1.5–1.75 mm in diameter.

Etymology

Deroceras: "neck-horn"; *panormitanum*: after Panormus, the classical name for Palermo, Italy, where the species was first discovered.

Selected References

Barker 1999, Giusti and Manganelli 1990, Hanna 1966, Kerney and Cameron 1979, Pilsbry 1948, Quick 1960, Rollo and Wellington 1975, Webb 1961.

Deroceras (Deroceras) reticulatum (Müller, 1774)   Grey Fieldslug

Synonym: *Limax reticulatum* Müller, 1774; *Agriolimax agrestis* of authors, not Linnaeus, 1758.

Description

This slug is 35–50 mm long, cream-coloured, greyish or slightly pinkish grey, usually with darker brown or grey flecks mostly between the tubercles. Mucus thick and sticky, becoming milky white when the animal is irritated. Sole of the foot is whitish. Intestinal caecum present. Colour photograph C-23. Genitalia: penis constricted at the middle; penial flagellum consists of 1–4 irregular, branched, knobby processes (indicated by the arrow in the illustration); stimulator – within

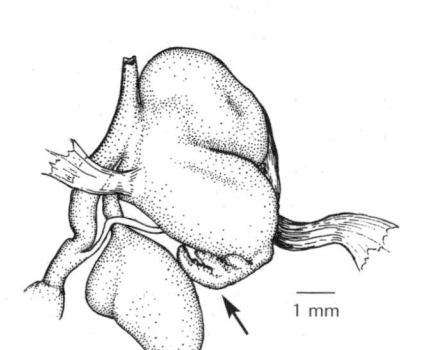

1 mm

Distal genitalia (the arrow points to the penial flagellum).

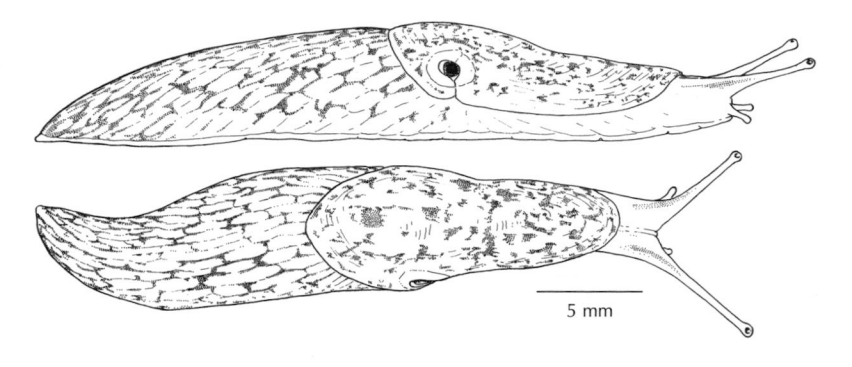

5 mm

the penis – conical. Illustrated in Barker 1999, Gittenberger et al. 1984, Kerney and Cameron 1979, Pilsbry 1948, Quick 1960 and Wiktor 2000.

This is the largest species of its genus in British Columbia, recognizable by its size, its milky white mucus and the structures of its penis.

Distribution
Probably native to western Europe; now widely introduced to many places worldwide. First reported in Victoria in the early 1890s as *Limax agrestis*, this species is now one of the most common and widespread of exotic slugs in British Columbia; it occurs as far north as the Queen Charlotte Islands, Mackenzie and the Bulkley Valley.

Natural History
Deroceras reticulatum is often abundant in gardens, roadsides, fields and other modified habitats. It is omnivorous, but since a large part of its diet includes green plants, it considered a serious agricultural and horticultural pest.

A slow moving species, *D. reticulatum* detects predacious carabid beetles by smell. When bitten by a beetle, this slug can lash its tail, discharge large amounts of mucus and flee; self-amputation, or autotomy, of the tip of the tail has also been observed in this species.

Reproduction is predominantly by cross-fertilization and occurs throughout the year. Animals live for one to two years.

Etymology
Deroceras: "neck-horn"; *reticulatum*: "netted", referring to the pattern on the skin surface of the tail.

Selected References

Barker 1999; Dodds et al. 1997; Getz 1959; Pakarinen 1994a, b; Pilsbry 1948; Rollo and Wellington 1975; Runham and Laryea 1968; South 1965; Stephenson 1979; Taylor 1891b; Wiktor 2000. (This is a small sample of the extensive body of literature available on this species.)

Family Arionidae

Ariolimax (*Ariolimax*) *columbianus*   Pacific Bananaslug (Gould *in* A. Binney, 1851)

Synonyms: *Limax columbianus* Gould *in* A. Binney, 1851; *Ariolimax columbianus maculata* Cockerell, 1891; *A. columbianus niger* Cockerell, 1891; *A. columbianus typica* Cockerell, 1891; *Aphallarion buttoni* Pilsbry & Vanatta, 1896; *Ariolimax steindachneri* Babor, 1900.

Description

This large slug grows to a length of about 260 mm. It is ochre yellow, olive-green, brownish green, reddish brown or white and often spotted with irregular black blotches that may coalesce and make the animal entirely black; see colour photographs C-16, C-17 and C-18. Mantle finely granular, often with a single, centrally placed dark spot. Pneumostome placed behind the mid-point of the mantle. Keel does not quite reach the mantle; in strongly contracted animals, it is distinctly wavy. Caudal mucus pore present. Genitalia: Mead 1943, Pilsbry 1948.

Ariolimax columbianus is distinguished by its large size, partially keeled tail and colouration. Juvenile slugs may be finely speckled.

Distribution
Southeastern Alaska to southern California; northern Idaho. In British Columbia, *Ariolimax columbianus* is a well-known, common species on the west sides of the Boundary, Coast and Cascade mountains and on coastal islands, including the Queen Charlotte Islands.

Natural History
We know more about the natural history of Bananaslugs than most of our other native species. They are common inhabitants of humid coastal forests, often seen making their way across a trail. They are renowned among biologists for their peculiar habit of gnawing off their partner's penis immediately following copulation. *Ariolimax columbianus* lays oval eggs (about 5 × 8 mm) in clutches in the soil from autumn to early spring; the eggs hatch in three to eight weeks. It eats green and dead vegetation, fungi, feces and dead slugs. Like many slugs, it shows evidence of homing behaviour.

Etymology
Ariolimax: a compound of *Arion* and *Limax*, two genera of Old World slugs; *columbianus*: after the Columbia River.

Remarks
Biologists recognize two subspecies of *Ariolimax columbianus*: a northern subspecies, *A. columbianus columbianus*; and a southern subspecies, *A. columbianus stramineus* Hemphill 1891. The ranges of the two subspecies meet in central California. They differ in their reproductive anatomy and external pigmentation.

Selected References
Dall 1905; Frest and Johannes 2001; Groves 1992; Ingram and Peterson 1947; Pilsbry 1948; Richter 1979, 1980; Rollo and Wellington 1975, 1981; Roth 2004; Westfall 1959.

COLOUR PHOTOGRAPHS

All photographs by Kristiina Ovaska
[except where noted].

C-1. *Anguispira kochi* (Banded Tigersnail) at Beauty Creek, Idaho. Page 79.

C-2. *Monadenia fidelis* (Pacific Sideband) in Goldstream Provincial Park, Vancouver Island. Page 159.

C-3. *Cepaea nemoralis* (Grovesnail) at Vancouver. Page 160.

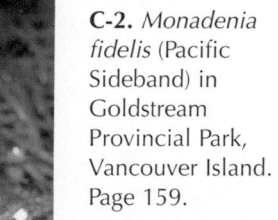

C-4. *Vespericola columbianus* (Northwest Hesperian) in Goldstream Provincial Park, Vancouver Island. Page 155.

C-5. This *V. columbianus* at Metchosin, Vancouver Island, lacks the hairy periostracum typical of most individuals. Page 155.

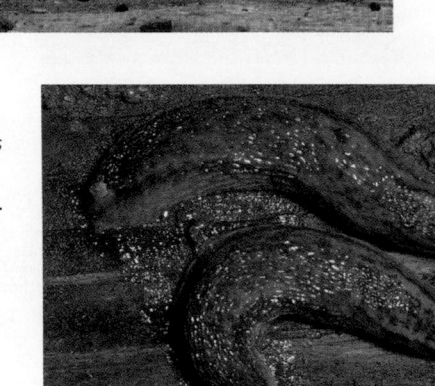

C-6. *Arion distinctus* (Darkface Arion) at Victoria. Page 132.

C-7. *A. circumscriptus* (Brown-banded Arion) at Victoria. Page 129.

C-8. *Arion rufus* (Chocolate Arion), brown form with orange foot fringe, at Prospect Lake, Vancouver Island. [Robert Forsyth photo.] Page 127.

C-9. *A. rufus,* black colour form, in Goldstream Provincial Park, Vancouver Island. [Robert Forsyth photo.] Page 127.

C-10. *A. subfuscus* (Dusky Arion) at Alice Arm, Observatory Inlet. [Robert Forsyth photo.] Page 135.

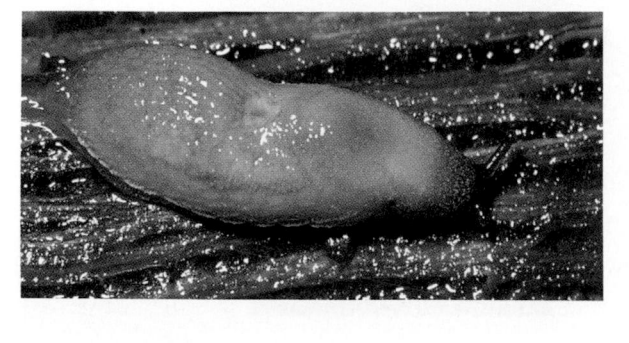

C-11. *A. intermedius* (Hedgehog Arion) at Victoria. Page 133.

C-12. *Boettgerilla pallens* (Wormslug) at Metchosin, Vancouver Island. Page 110.

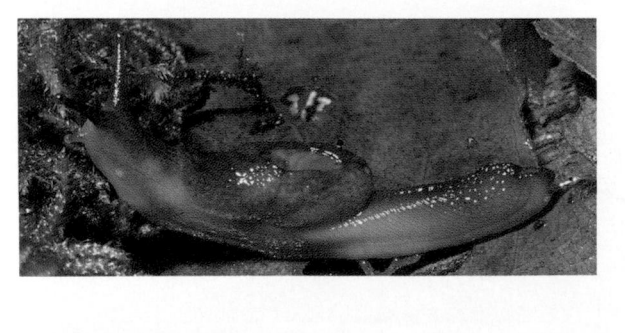

C-13. *Hemphillia dromedarius* (Dromedary Jumping-slug) near Duncan, Vancouver Island. Page 137.

C-14. *H. camelus* (Pale Jumping-slug) in Pend Oreille County, Washington. Page 136.

C-15. *H. glandulosa* (Warty Jumping-slug) on Mt Brenton, Vancouver Island. Page 139.

C-16. *Ariolimax columbianus* (Pacific Bananaslug) in Goldstream Provincial Park, Vancouver Island. Page 123.

C-17. *A. columbianus*, light colour form, near Port Clements, Graham Island. Page 123.

C-18. *A. columbianus*, dark colour form, near Port Clements, Graham Island. Page 123.

C-19. *Haplotrema vancouverense* (Robust Lancetooth) in Goldstream Provincial Park, Vancouver Island. Page 72.

C-20. *Oxychilus draparnaudi* (Dark-bodied Glass-snail) at Victoria. Page 107.

C-21. *Deroceras laeve* (Meadow Slug) at Victoria. Page 118.

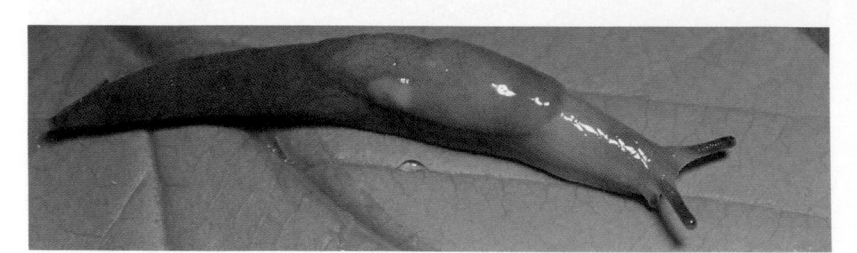

C-22. *D. panormitanum* (Longneck Fieldslug) at Victoria. Page 120.

C-23. *D. reticulatum* (Grey Fieldslug) at Morfee Lakes, near Mackenzie. [Robert Forsyth photo.] Page 121.

C-24. *Oxychilus alliarius* (Garlic Glass-snail) at Vancouver. Page 104.

C-25. *Prophysaon coeruleum* (Blue-grey Taildropper) at Metchosin, Vancouver Island. Page 144.

C-26. *P. andersoni* (Reticulate Taildropper) at Victoria. Page 142.

C-27 (below). *P. foliolatum* (Yellow-bordered Taildropper) in the Tsitika River valley, Vancouver Island. Page 145.

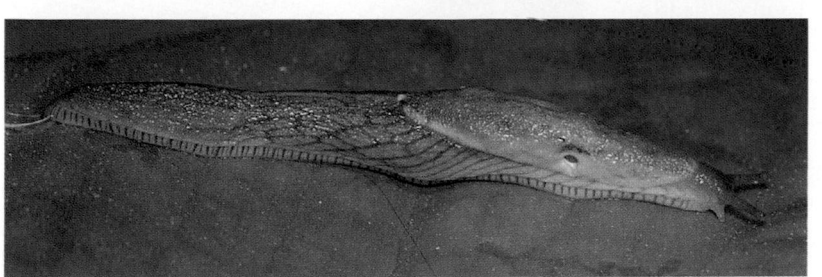

C-28. *P. vanattae* (Scarletback Taildropper) at Kennedy Lake, Vancouver Island. Page 141.

C-29. *Allogona townsendiana* (Oregon Forest-snail) at Langley. Page 149.

C-30. *Limax maximus* (Giant Garden-slug) near Duncan, Vancouver Island. Page 113.

C-31. *Lehmannia valentiana* (Three-band Gardenslug) at Victoria. Page 115.

C-32. *Limacus flavus* (Yellow Gardenslug) at Central Saanich, Vancouver Island. [Kelly Sendall photo.] Page 112.

C-33. *Ancotrema hybridum* (Oregon Lancetooth) near Port Clements, Graham Island. Page 70.

Magnipelta mycophaga Pilsbry, 1953   Magnum Mantleslug

Description
A large slug about 65 mm long, with a light tan colour and black spots. The mantle covers most of the dorsal surface; it is smooth and tan coloured with an irregular black stripe on either side and irregular black spotting; the anterior quarter is free from the head. Pneumostome slightly posterior to the midline on the right side of the mantle. No keel on the tail. Genitalia: Pilsbry and Brunson 1954, Webb and Russell 1977.

The extremely large mantle is distinctive and it is unlikely that this slug, the only member of its genus, would be mistaken for another.

Distribution
Northeast Washington, east through the Idaho Panhandle to western Montana, and north into southeastern British Columbia.

Natural History
This species lives in cool, moist coniferous forests under moist logs and pieces of bark, and in depressions in moist earth and talus. When threatened, *Magnipelta mycophaga* spreads its large mantle in wing-like fashion, sometimes curling back the anterior part. Its diet includes fungi.

Etymology

Magnipelta: "large shield", referring to the huge mantle; *mycophaga*: "fungus eating", because the first specimen found was feeding on fungus.

Selected References

Brunson and Kevern 1964, Forsyth 2004, Frest and Johannes 2001, Pilsbry 1953, Pilsbry and Brunson 1954.

Key to species of *Arion*

Species of *Arion* can be recognized by the granular mantle, the forward placed pneumostome, presence of a caudal mucus pore and coarse tubercles on the tail. But differentiating species in this genus may be difficult because there are groups of very similar species (complexes); they are best distinguished by differences in their sexually mature reproductive systems.

The following key applies to adult slugs only, because the diagnostic characters of the juvenile reproductive system may not be fully developed and the body pigmentation of juveniles may differ from adults.

1a   Length to about 15 mm, tubercles on the back having soft points, giving the contracted animal a prickly appearance; body yellowish grey with head darker and lateral bands usually faint .... *Arion intermedius*
1b   Adults longer than 20 mm, without a "prickly" appearance when contracted; variously pigmented 2
2a   Contracted body bell-shaped in cross-section (A) 3
      Contracted body not bell-shaped in cross-section (B)
      .... *A. distinctus*

3a   Large, greater than 70 mm (up to nearly 200 mm) long, coarse tubercles; usually not banded; often twists and rocks when disturbed .... *A. rufus*
3b   Less than 70 mm in length; fine tubercles; usually banded; does not rock or twist when disturbed 4

4a   General body colour reddish brown, rusty orange or occasionally yellow; yellow or orange body mucus; cannot contract into a hemispherical shape (viewed from the side) .... *A. subfuscus*

4b   General body colour light grey or brownish grey, or with yellow on it; colourless body mucus; contracts into a hemispherical shape 5

5a   Body light grey with a yellow or cream tinge and usually with a yellowish band below the lateral bands (mantle not dotted).. *A. fasciatus*

5b   Body light grey, without the yellowish tinge or band below the lateral bands 6

6a   Lateral bands more or less blurred below; mantle has dark dots; distal third of the epiphallus heavily pigmented with dark dots .... *A. circumscriptus*

6b   Body pale grey, fading to white on the sides, lateral bands well contrasted on lower edge; mantle not dotted; epiphallus unpigmented or only slightly pigmented............... *A. silvaticus*

Arion (Arion) rufus (Linnaeus, 1758)   **Chocolate Arion**
Synonym: *Limax rufus* Linnaeus, 1758; *Arion ater* of authors, not Linnaeus, 1758.

Description
A large slug that grows to about 180 mm long. It comes in various shades of brown to black or orange; see colour photographs C-8 and C-9. juveniles display a broader range of colour than adults, and adults do not have lateral bands. *Arion rufus* can contract its body into a bell-shape, viewed from the side. Foot fringe red, orange, yellow or grey, with vertical black bars. Sole of the foot whitish or greyish white, with broad blackish bands on either side of a light central area, or all black. Mucus colourless. Genitalia: Cain and Williamson 1958, Quick 1960, Kerney and Cameron 1979.

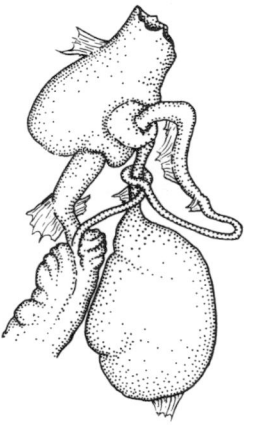

Distal genitalia.

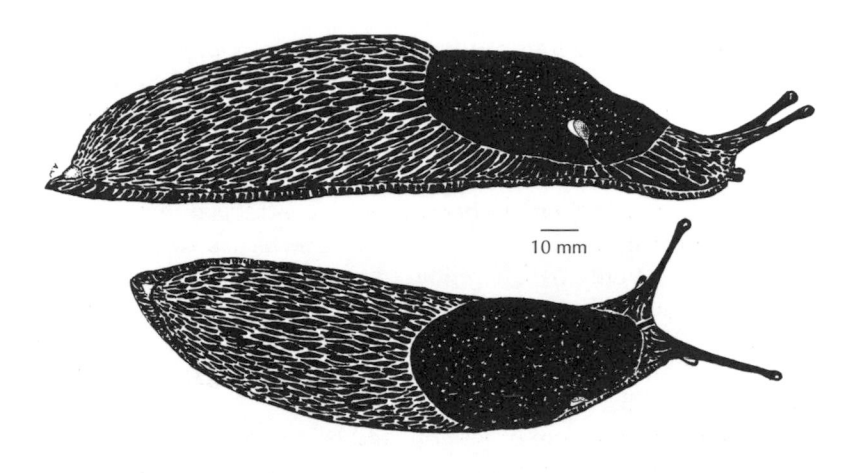

10 mm

Juveniles of this species are yellow and often banded and could be confused with other species of *Arion*, particularly *Arion intermedius*. Aside from *A. ater* (see page 147), which is very similar to *A. rufus*, adult slugs can be easily recognized by their colour, large size and coarse granulation.

Distribution
Western and Central Europe; probably introduced to many places worldwide and often misidentified as *Arion ater*. *A. rufus* is one of the more familiar slugs in southern British Columbia, and is also introduced to the Queen Charlotte Islands.

Natural History
These slugs live in gardens, fields, campgrounds and other disturbed sites, and apparently are easily transported to new sites. When disturbed, *Arion rufus* will contract and then often twist and rock from side to side. It is omnivorous, eating live plants and dead plant material, fungi, animal faeces and carrion. Mating takes place through summer and early autumn.

Etymology
Arion: the name of a Greek poet and musician; *rufus*: "red" or "reddish".

Selected References
Gittenberger et al. 1984; Quick 1947, 1960; Rollo and Wellington 1975; Roth and Pearce 1984.

Arion (*Carinarion*) *circumscriptus*   Brown-banded Arion
Johnston, 1828

Synonym: *Arion fasciatus* of authors, in part, not Nilsson, 1823.

Description

This slug is 30–40 mm long, pale grey above, fading to whitish on the sides. Mantle grey, marked with small dark spots. Colour photograph C-7. Genitalia: atrium relatively large, at least twice as long as broad; epiphallus small and narrow (distal part about as wide as oviduct); epiphallus with dark speckles on its distal third; base of the bursa copulatrix duct not strongly swollen. See Kerney and Cameron 1979, Gittenberger et al. 1984.

Slugs of the subgenus *Carinarion* are medium-sized and have a false keel (especially in juveniles), which consists of a central row of pale, enlarged tubercles. The body, when contracted, is bell-shaped in cross-section and greyish with darker lateral bands. The foot fringe is pale, the sole of the foot is white. *Carinarion* species are most clearly distinguished by their reproductive anatomy, but the presence of small dark spots on the mantle of *A. circumscriptus* is a good feature for its recognition from other *Carinarion*.

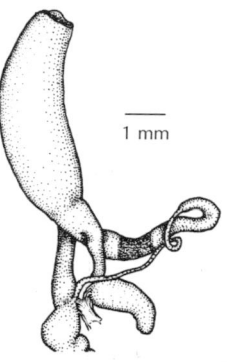

1 mm

Distal genitalia.

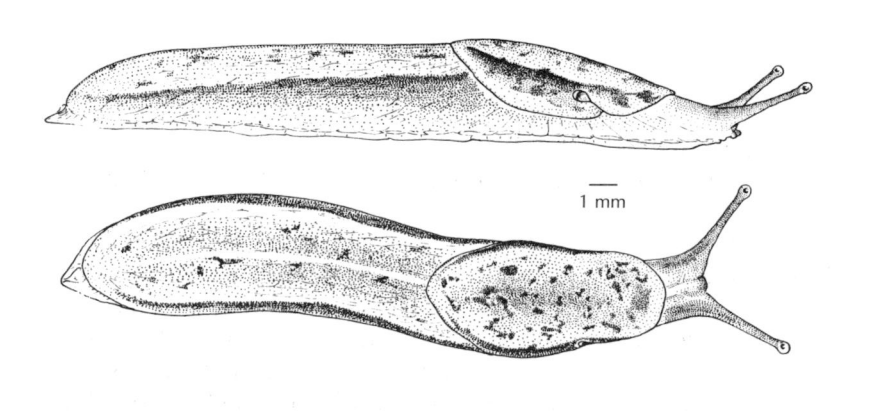

1 mm

Distribution
Europe, but introduced to North America. It is common in many populated places throughout British Columbia. In the north-central part of the province, *Arion circumscriptus* is the most common species of *Arion*.

Natural History
This species is often found in large numbers in gardens, disturbed sites and woods close to human settlements or activities.

Etymology
Arion: the name of a Greek poet and musician; *Carinarion*: "keeled *Arion*", a reference to the false keel; *circumscriptus*: "written around", perhaps in reference to the lateral band around the pneumostome.

Remarks
Genetic and morphological studies have placed some doubt on the status of the *Carinarion* species. For example, *Arion fasciatus* and *A. silvaticus* cannot be distinguished in some geographical areas and appear to cross with each other. There is also evidence that diet can affect skin pigmentation in species of *Carinarion*.

Selected References
Backeljau et al. 1997; Getz 1959; Jordaens et al. 2000, 2001; Kerney 1999; Kerney and Cameron 1979; Pilsbry 1948; Rollo and Wellington 1975.

Arion (Carinarion) silvaticus   **Forest Arion**
Lohmander, 1937
Synonym: *Arion fasciatus* of authors, in part, not Nilsson, 1823.

Description
This slug is 30–40 mm long, pale grey above, fading to whitish on the sides and paler below the lateral bands, which are broad and dark. Mantle has no dark spots. Genitalia: intermediate between *Arion circumscriptus* and *A. fasciatus*. Atrium usually as long and narrow as that of *A. circumscriptus*; epiphallus medium-sized and not pigmented; proximal portion of epiphallus and oviduct about

equal in width; bursa copulatrix duct may or may not have a distal swollen section. See Kerney and Cameron 1979, Gittenberger et al. 1984.

Arion silvaticus can be distinguished from *A. fasciatus* with certainty only by differences in genitalia; it differs from *A. circumscriptus* by its lack of mantle speckles as well as anatomical differences.

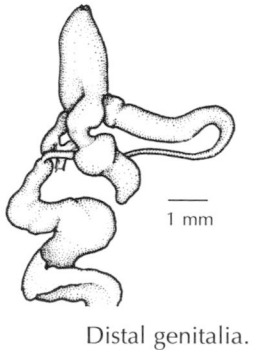

1 mm

Distribution
Europe. *Arion silvaticus* was found near Vancouver, but it appears to be scarce and not as widespread as *A. circumscriptus*. Also found in San Francisco, California.

Distal genitalia.

Natural History
Arion silvaticus lives in moist places under rocks, vegetation and debris in gardens, parks and modified habitats.

Etymology
Arion: the name of a Greek poet and musician; *Carinarion*: "keeled *Arion*", a reference to the false keel; *silvaticus*: Latin, "of the woods".

Remarks
See Remarks under *Arion circumscriptus* (page 130).

Selected References
Backeljau et al. 1997, Jordaens et al. 2000, Kerney 1999, Kerney and Cameron 1979, Rollo and Wellington 1975, Roth 1986.

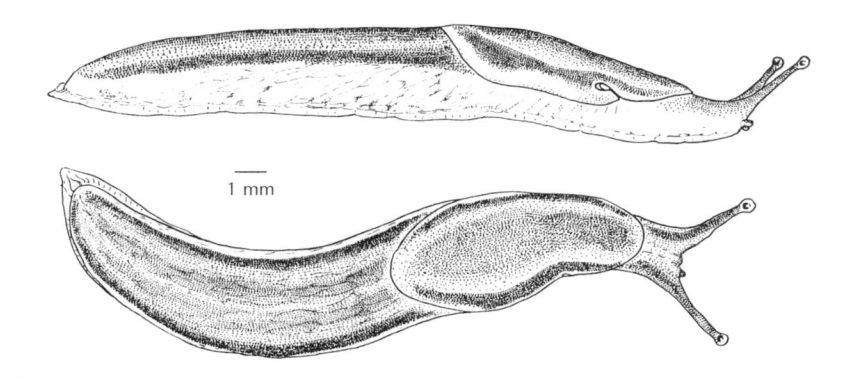

1 mm

Arion (*Kobeltia*) *distinctus* Mabille, 1868   Darkface Arion

Synonym: *Arion hortensis* of authors in part, not Férussac, 1819; *Arion hortensis* form 'A' of authors.

Description

This slug is 25–30 mm long (rarely larger), dark grey or bluish grey, sometimes with a brownish tinge, and has dark lateral bands, the right band usually encompassing the pneumostome. Tentacles bluish grey. Contracted animals are rounded rather than bell-shaped in cross-section. No keel. Foot sole pale yellow or orange. Mucus yellow-orange. Colour photograph C-6. Genitalia: Davies 1977, 1979; Gittenberger et al. 1983.

Its pigmentation and bright yellow-orange mucus distinguishes *Arion distinctus* from other confirmed *Arion* species in this book, but see Remarks.

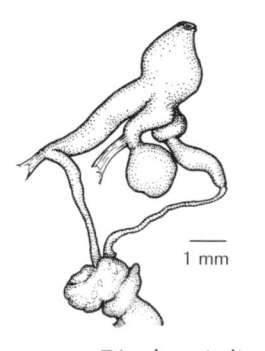

1 mm

Distal genitalia.

Distribution

Native to western and southern Europe, and introduced to North America, New Zealand and probably elsewhere but likely confused with *A. hortensis*. In British Columbia, *A. distinctus* is recorded from southern Vancouver Island and the Lower Mainland and likely occurs elsewhere.

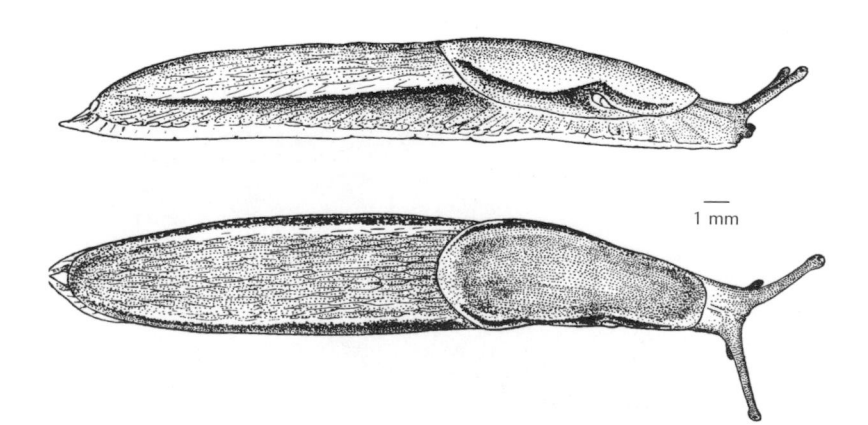

1 mm

Natural History
This species lives in moist places in gardens and parks.

Etymology
Arion: the name of a Greek poet and musician; *Kobeltia*: named for Wilhelm Kobelt (1840–1916), German malacologist; *distinctus*: "distinct", although this species' distinctness was not shown until recently.

Remarks
Arion distinctus is one of three similar, closely related species, collectively called the *A. hortensis* aggregate: *Arion hortensis* Férussac, 1819; *A. owenii* Davies, 1979; and *A. distinctus*. These species are well defined by aspects of their anatomy, behaviour and natural history. *A. distinctus* and *A. hortensis* are only reliably distinguishable by genital anatomy, specifically the form of an epiphallus structure, which is located internally at the junction of the epiphallus and atrium.

Selected References
Backeljau 1985; Backeljau and Van Beeck 1986; Davies 1977, 1979; Kerney 1999; Kerney and Cameron 1979; Roth 1982a.

Arion (*Kobeltia*) *intermedius*   Hedgehog Arion
Normand, 1852

Description
A small slug, 15–20 mm long, usually yellowish-grey with darker grey head and tentacles. Lateral bands pale or absent. When contracted, the body is not bell-shaped in cross-section, but the tubercles form sharp, translucent points. No keel. Foot fringe pale and narrow. Sole of foot yellowish-grey to pale orange. Mucus yellowish or orange-yellow. Colour photograph C-11. Genitalia: Gittenberger et al. 1984, Quick 1960.

    Arion intermedius is distinguished by its small size, yellowish pigmentation, yellowish mucus and the prominently pointed tubercles when contracted. Because of its size, it can be confused with juveniles of other *Arion* species, particularly *A. rufus*.

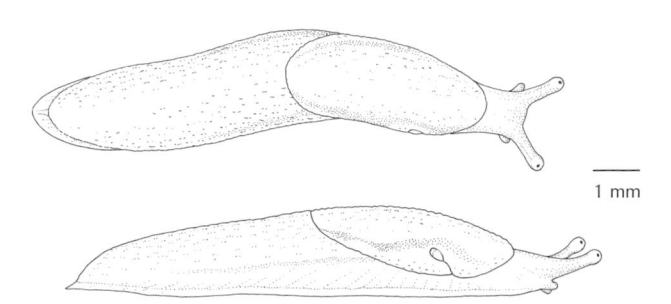

1 mm

Distribution
Native to central and northern Europe; introduced to North America, Polynesia, New Zealand and North Africa. In British Columbia, *Arion intermedius* has been found in Greater Vancouver, on the Queen Charlotte Islands and on southern Vancouver Island.

Natural History
Arion intermedius feeds on green plants and fungi. Reproduction is predominantly by self-fertilization although outcrossing does occasionally occur. In British Columbia, this species lives in fields, grassy roadsides and mature gardens. In Europe, it is also found in woods .

Etymology
Arion: the name of a Greek poet and musician; *Kobeltia*: named for German malacologist Wilhelm Kobelt (1840–1916); *intermedius*: "intermediate", but the significance of this is unknown.

Selected References
Barker 1999, Davies 1977, Garrido et al. 1995b, Kerney 1999, Kerney and Cameron 1979, Pilsbry 1948, Quick 1960, Reise et al. 2001, Rollo and Wellington 1975.

Arion (Mesarion) subfuscus (Draparnaud, 1805)

Dusky Arion

Synonym: *Limax subfuscus* Draparnaud, 1805.

Description

This slug is 35–70 mm long and orange-yellow to brown, usually with prominent brown lateral bands. Its contracted body is not bell-shaped in cross-section. Foot fringe pale coloured, marked with darker vertical stripes. Sole of the foot is cream coloured; mucus yellowish or orange-yellow. Colour photograph C-10. Genitalia: Gittenberger et al. 1984, Quick 1960.

Arion subfuscus is distinguished by its size, its orange pigmentation, its inability to fully contract into a hemispherical form and its anatomy.

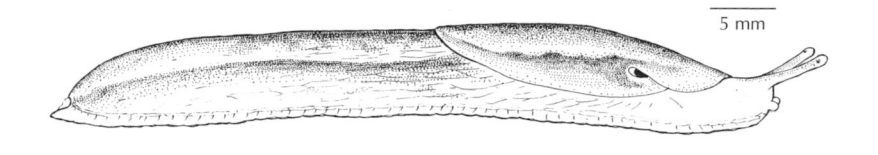

5 mm

Distribution

Native to Europe; introduced to North America. In British Columbia, it is generally distributed near human settlements from the south to at least the Queen Charlotte Islands, Observatory Inlet and the Bulkley Valley.

Natural History

These slugs live along roadsides and in gardens, and woods adjacent to human settlement. They eat fungi, microscopic algae, green plants and carrion.

Etymology

Arion: the name of a Greek poet and musician; *Mesarion*: "middle *Arion*"; *subfuscus*: "somewhat brown".

Remarks

Recent studies into the genetics of these slugs have established that *Arion subfuscus* is a complex of at least two species that can hybridize.

Selected References

Backeljau et al. 1996, Garrido et al. 1995a, Kerney 1999, Kerney and Cameron 1979, Pilsbry 1948, Pinceel et al. 2002, Rollo and Wellington 1975, Roth and Pearce 1984.

Key to species of *Hemphillia*

1a Small, to about 30 mm long. Visceral lump and mantle cover half the body length or more and have many small papillae .... *Hemphillia glandulosa*

1b Animal can measure up to about 60 mm long. Visceral hump and mantle cover less than half the body length and have only a few scattered papillae 2

2a Prominent caudal horn (a fleshy protuberance at the end of the tail above the caudal mucus pore). Penis has an accessory sac .... *H. dromedarius*

2b Caudal horn absent. Penis has no accessory sac .... *H. camelus*

Hemphillia camelus   **Pale Jumping-slug**
Pilsbry & Vanatta, 1897

Description

This slug grows to about 55 mm long and is pale yellowish brown or ashy grey, speckled and spotted with black or greyish brown. Mantle wrinkled, without papillae; spots on the mantle coalesce into lateral bands. Caudal horn absent. Tail not raised into a high keel. Colour version of photograph C-14. Genitalia: Pilsbry 1948.

This species is similar to *Hemphillia dromedarius*, but is paler and has no caudal horn.

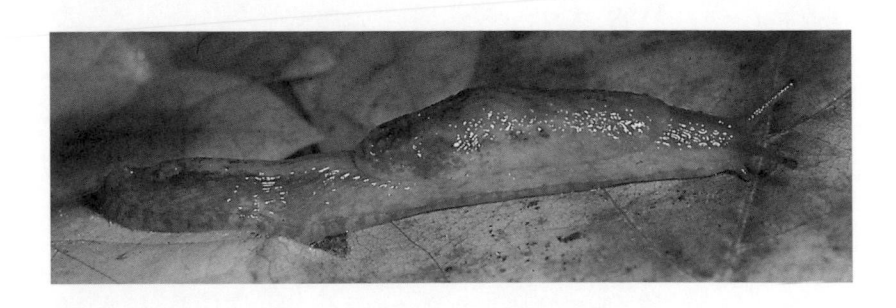

Distribution

Southeast British Columbia, Alberta, northeastern Washington, Idaho and northwestern Montana. In British Columbia, this is our most widespread *Hemphillia*, occurring north to East Barriere Lake, Kamloops, Glacier National Park and near Golden.

Natural History

Hemphillia camelus occurs in dry to moist coniferous forests where it lives on and around mossy stumps, rocks and logs and in leaf litter. As a defence against predators, all species of *Hemphillia* curl their tail around to the side of their body, and when irritated, may thrash from side to side and flip themselves away from danger.

Etymology

Hemphillia: named for the malacologist Henry Hemphill (1830–1914), who collected extensively throughout the western United States; *camelus*: "camel", for the humped back.

Selected References

Frest and Johannes 2001, La Rocque 1953, Pilsbry 1948.

Hemphillia dromedarius   Dromedary Jumping-slug
Branson, 1972

Synonym: *Hemphillia malonei* of Hanham 1926, not Pilsbry, 1917.

Description

This slug grows to about 60 mm long and is grey with cream mottling. Mantle and visceral pouch are nearly black to yellowish, densely spotted with bluish black and grey; mantle wrinkled, with a few well-spaced papillae. Caudal horn prominent. Foot sole cream coloured, pale yellow or yellowish orange. Tail not raised into a high keel. Genitalia: Ovaska et al. 2002.

Hemphillia dromedarius is similar to *H. camelus*, which is paler and does not have a caudal horn.

Distribution

Southern Vancouver Island to the Cascade Range and Olympic Peninsula, Washington; in British Columbia, it is apparently restricted to Vancouver Island.

See the colour version of this photograph, C-13.

Natural History
This slug is typical of old-growth and older second-growth conifer-ous forests where it occurs in low densities. It finds shelter under logs, rocks and vegetation. The eggs are semi-opaque and oval and are laid in clusters in moist rotting wood. As a defence against predators, all species of *Hemphillia* curl their tail around to the side of their body, and when irritated, may thrash from side to side and flip themselves away from danger.

Etymology
Hemphillia: named for the malacologist Henry Hemphill (1830–1914), who collected extensively throughout the western United States; *dromedarius*: named for the Arabian or one-humped camel, *Camelus dromedarius*.

Remarks
As of 2003, *Hemphillia dromedarius* has been on the Canadian list of Species at Risk as "Threatened" because of rarity and association with old-growth and older second-growth forests.

Selected References
Branson 1972, 1980; Ovaska et al. 2002.

Hemphillia glandulosa
Bland & W.G. Binney, 1873

Warty Jumping-slug

Description
This slug's body is thick and about 30 mm long; it is light brownish and has a bluish grey head and tentacles. Mantle and visceral hump cover half or more of the body length; mantle streaked and dark spotted; visceral pouch covered with numerous, densely packed papillae. Dorsal keel high and narrow. Caudal horn large. Genitalia: Pilsbry 1948.

This species differs from other *Hemphillia* by its short, thick body, relatively large mantle and visceral hump, and the presence of closely spaced papillae.

See the colour version of this photograph, C-15.

Distribution
Nanaimo River, Vancouver Island to northwest Oregon. In British Columbia, this species has been found only on Vancouver Island.

Natural History
Hemphillia glandulosa occurs sporadically in coniferous, deciduous and mixed-wood forests, where it may occur at higher densities than *H. dromedarius*. It is often found under logs, ferns and other vegetation, and in litter. As a defence against predators, all species of *Hemphillia* curl their tail around to the side of their body, and when irritated, may thrash from side to side and flip themselves away from danger.

Etymology
Hemphillia: named for the malacologist Henry Hemphill (1830–1914), who collected extensively throughout the western United States; *glandulosa*: "with small glands" possibly a reference to the papillae on the mantle.

Remarks
As of 2003, *Hemphillia glandulosa* has been on the Canadian list of Species at Risk as "Special Concern". Habitat loss, scattered populations and a small range in Canada leaves it susceptible to natural and human disturbances.

Selected References
Branson 1977, Taylor 1900, Webb 1961.

Key to species of *Prophysaon*

1a   Epiphallus very long and slender, except for an abruptly enlarged sausage-shaped, banana-shaped or globular thick-walled portion near where it connects to the penis................. 2

1b   Epiphallus moderately long and slender with no an abruptly enlarged portion .... *Prophysaon vanattae*

2a   Muscular structure of the epiphallus globular. Animal blue-grey; body length to 45 mm...................................... *P. coeruleum*

2b   Muscular structure of the epiphallus elongate. Animal variously pigmented but not blue-grey; body length of adults generally greater than 45 mm 3

3a   Autotomy zone on the tail at about a quarter the length of the entire animal; foot fringe usually lacks dark vertical bars (sometimes they are present but faint); body length of adults up to about 60 mm..................................... *P. andersonii*

3b   Autotomy zone on the tail at about a third of the length of the entire animal; foot fringe usually has dark vertical bars (but sometimes faint); adults can reach up to about 120 mm long (but can be as small as about 30 mm) .... *P. foliolatum*

Prophysaon (Mimetarion)   Scarletback Taildropper
vanattae **Pilsbry, 1948**
Synonym: ?*Prophysaon obscurum* Cockerell, 1890.

Description
This slug is 25–40 mm long (sometimes longer). Its colour varies: whitish buff, bluish grey or red on the back; grey-buff on the sides; and usually with two conspicuous lateral bands running back along the tail from the mantle, defining a wedge-shaped, lighter dorsal area that may enclose another darker area. Mantle buff or reddish with two dark lateral bands (in some these band are faint) and scattered black maculations or marbling. Sole dark grey. Colour photograph C-28. Genitalia: epiphallus tapers and its inner wall has two longitudinal ridges of roughly equal size; see Pilsbry 1948.

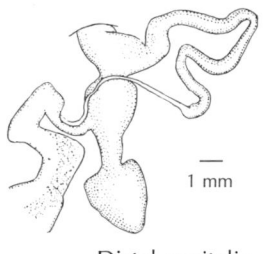

1 mm

Distal genitalia.

Distribution
Vancouver Island and the Cascade Mountains of British Columbia, to northwestern Oregon.

Natural History
Prophysaon vanattae is mainly arboreal, living on the moss-covered trunks and branches of trees and shrubs in coastal mixed-wood forests.

Etymology

Prophysaon: "forward breathing", referring to the placement of the pneumostome in the front part of the mantle; *mimetarion*: compound meaning "mimic-*Arion*"; *vanattae*: after Edward G. Vanatta.

Remarks

Prophysaon obscurum may be a dark colour form of *P. vanattae*.

Selected References

Branson 1977, Cameron 1986, Pilsbry 1948.

Prophysaon (Prophysaon)   **Reticulate Taildropper**
andersonii **(J.G. Cooper, 1872)**
Synonyms: *Arion andersonii* J.G. Cooper, 1872; *Prophysaon hemphilli*
Bland & W.G. Binney, 1873; *P. pacificum* Cockerell, 1890; *P. flavum*
Cockerell, 1890; *P. andersoni marmoratum* Cockerell, 1892.

Description

This slug grows to about 60 mm long. It is pale brownish, reddish grey or yellowish, clouded with darker tones and has a diamond-mesh pattern on the back. Mantle often paler and finely granular, usually with a pair of dark lateral bands. Pneumostome close to the middle of the mantle or in the anterior half. No keel. Foot fringe pale and usually without dark vertical

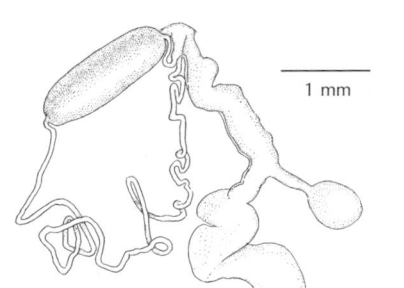

1 mm

Distal genitalia.

bars (though sometimes there are faint bars). Body mucus lemon yellow to orange. Sole brilliant white. Animal tadpole-shaped when contracted. Colour photograph C-26. Genitalia: Pilsbry 1948.

Prophysaon andersonii is smaller than *P. foliolatum* and its autotomy zone is further back (i.e., the self-amputated portion is relatively shorter). In *P. andersoni* the enlarged, muscular portion of the epiphallus may be relatively shorter and straighter, but in both

species the character of this structure is variable. *P. vanattae* has black bands on the mantle and tail, and the body often has orange pigment, and has a gradually tapering epiphallus (typical of its subgenus) without a distinctly swollen cylindrical portion.

Distribution
Aleutian Islands, Alaska to California and possibly east to Idaho. In British Columbia it has been found south of Observatory Inlet on the north coast, the Bulkley Valley, Fort St James and Mackenzie inland.

Natural History
Prophysaon andersonii occurs in woods and mildly disturbed sites, where it finds shelter in moist places under logs, leaf litter, rocks and vegetation. It reaches maturity and mates in late summer and fall. As a defence against predators, this and other species of *Prophysaon* are capable of autotomy of the tail, and a diagonal constriction on the tail usually marks the site where amputation occurs.

Etymology
Prophysaon: "forward breathing", referring to the placement of the pneumostome in the front half of the mantle; *andersonii*: after Dr C.L. Anderson of Santa Cruz, California, an avid naturalist.

Selected References
Hand and Ingram 1950, Pilsbry 1948, Rollo and Wellington 1975.

Prophysaon (Prophysaon) *coeruleum* Cockerell, 1890

Blue-grey Taildropper

Description
This small slug grows to 45 mm long. It is blue-grey, but paler on the sides and behind the head and it has light flecks on the tail and mantle (but no bands on the mantle). Tail has distinct longitudinal grooves, occasionally coalescing or joining with transverse grooves. Foot fringe narrow with a distinct border above. Sole white. Colour photograph C-25. Genitalia: epiphallus has a bulbous muscular swelling; see Pilsbry 1948, Ovaska et al. in press.

This species' blue-grey colour and pattern of reticulations on the tail distinguishes it from all other slugs.

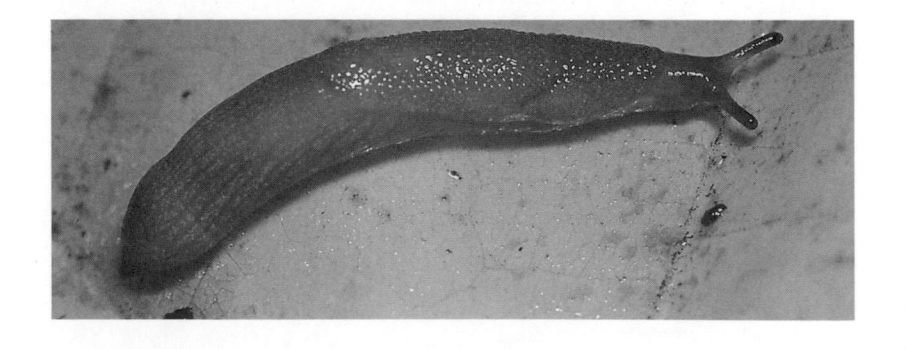

Distal genitalia.

Distribution
Southern Vancouver Island, British Columbia, and the Puget Trough, Washington, south to Northern California and east to Idaho.

Natural History
This species favours moist, mature or old-growth coniferous forests with a deciduous component (Bigleaf Maple or Trembling Aspen) and downed wood. *Prophysaon coeruleum* usually eats fungi, but also consumes green plant material, roots, lichens and moulds.

Etymology
Prophysaon: "forward breathing", referring to the placement of the pneumostome in the front part of the mantle; *coeruleum*, "blue".

Selected References
Branson and Branson 1984, McGraw et al. 2002, Ovaska et al. in press.

Prophysaon (Prophysaon)　Yellow-bordered Taildropper *foliolatum* (Gould *in* A. Binney, 1851)
Synonym: *Arion foliolatus* Gould *in* A. Binney, 1851.

Description
This large slug is usually longer than 100 mm, though sometimes much smaller; it is yellowish or brownish grey with dark reticulations on the tail. Foot fringe greyish, usually with dark vertical stripes (though sometimes very faint). Mantle edge usually yellow; mantle has dark, irregular mottling and obscure bands (formed by coalesced spots); dark bands interrupt the yellow mantle border, often leaving a yellow spot at the posterior end of the mantle. Tail has a light dorsal stripe. Sole cream coloured; mucus yellowish. Animal tadpole-shaped when contracted.
Genitalia: Pilsbry 1948.

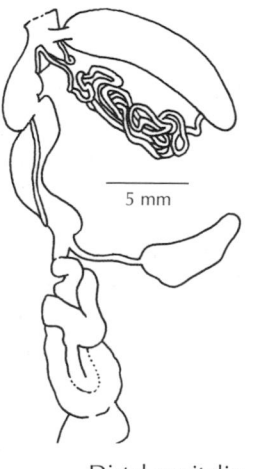

5 mm

Distal genitalia.

　　Prophysaon foliolatum is the largest species in the genus and has a distinctive bright yellow edge on its mantle and a yellowish sole, but its body colouration is variable. The oblique groove on the tail – the site of self-amputation – is usually strongly demarked, often by a dark line on the sole, and compared to *P. andersonii*, which this species most closely resembles, the groove is further forward on the body (resulting in a relatively longer section of the tail that can be self amputated). The relative length and curvature of the enlarged part of the epiphallus, as well as the relative length of the slender part, do not work to distinguish *P. foliolatum* from *P. andersonii*. (The relative

See the colour version of this photograph, C-27.

length of these structures in both species overlap. In *P. andersonii,* the enlarged part of the epiphallus can be straight or curved; in *P. foliolatum* it is also curved.) See also the comparison under the description of *P. andersonii.*

Distribution
Queen Charlotte Islands, British Columbia, to southwestern Oregon.

Natural History
Prophysaon foliolatum lives in wet, mainly coniferous, coastal forests. It is often encountered on the ground, on woody debris and vegetation (such as Skunk Cabbage and Devil's Club). These slugs feed on fungi and green plant material, mature and mate in the fall and can "drop" their tails as do other *Prophysaon* species.

We know more about the process of tail autotomy in *P. foliolatum* than any other *Prophysaon.* In a study by Deyrup-Olsen, Martin and Paine, this slug dropped its tail surprisingly quickly, in two to five seconds. After stimulation by a predator, such as the biting of a beetle, the autotomy zone contracted, the tail section swelled slightly, the body in front of the autotomy zone secreted large amounts of sticky mucus, and then the tail was completely severed. These slugs dropped their tails only when stimulated at or behind the autotomy zone, suggesting that the sticky mucus produced at the front discourages predators and redirects them to the less sticky tail section.

On Cleland Island, off the west coast of Vancouver Island, *P. foliolatum* has a unique association with seabirds in a peculiar habitat. There, it finds shelter in the nesting burrows of Leach's Storm Petrels and under driftwood.

Etymology
Prophysaon: "forward breathing", referring to the placement of the pneumostome in the front part of the mantle; *foliolatum:* "leafed", in reference to the veined pattern of reticulation on the tail.

Selected References
Branson and Branson 1984, Campbell and Stirling 1968, Deyrup-Olsen et al. 1986, Forsyth 2001a, Plisetskaya and Deyrup-Olsen 1987.

Unconfirmed species of Arionidae

Arion ater Linnaeus, 1758   Black Arion
This slug has been reported several times in British Columbia , but these records have not been confirmed. *Arion ater* is very similar to *A. rufus* – some malacologists consider the two species synonyms or subspecies. They are indistinguishable by external and behavioural characters, and can only be differentiated by reproductive anatomy: the atrium of *A. ater* is more slender than that of *A. rufus,* and its proximal portion is smaller and shorter than the distal part, which is the opposite in *A. rufus.*
Selected references: Carl and Guiguet 1972; Branson 1969; Cain and Williamson 1958; Kerney and Cameron 1979; Quick 1947, 1960; Rollo and Wellington 1975.

Arion (Carinarion) fasciatus   Orange-banded Arion (Nilsson, 1823)
Older records of *Arion fasciatus* may be correct, but most or all might be *A. circumscriptus,* the common member of this group in British Columbia. *A. fasciatus* is distinguished anatomically from other *Carinarion* by the relatively smaller atrium, which is more-or-less conical, the large, unpigmented epiphallus, which is distally at least twice as broad as the oviduct, and the distally swollen bursa copulatrix duct. The animal is usually light grey with a yellowish orange band below the lateral bands and no dark speckling on the mantle.
Selected references: Gittenberger et al. 1983, Kerney and Cameron 1979.

Arion (*Kobeltia*) *hortensis* Férussac, 1819   **Garden Arion**
Records of *Arion hortensis* from the Vancouver area predate its separation into three distinct species. The only reliable distinction
between two of these species – *A. distinctus* and *A. hortensis* – is by
the form of the epiphallus structure.
Selected references: Davies 1977, 1979; Rollo and Wellington 1975.

Family Polygyridae

Allogona (*Dysmedoma*) *ptychophora*   **Idaho Forestsnail**
(A.D. Brown, 1870)
Synonyms: *Mesodon townsendiana minor* Tryon, 1867; *Helix ptychophora* A.D. Brown, 1870; *Mesodon ptychophorus major* W.G. Binney,
1886; *H. ptychophorus castaneus* Hemphill, 1890.

Description
Shell 19–24 mm wide, heliciform and light
brown or straw yellow, with fine incremental striae and irregular, lighter-coloured,
wrinkle-like axial riblets that are strongest
near the suture. Exceedingly fine, wavy
spiral striae also present throughout the
shell; dimpled sculpture rarely present.
Periostracum sometimes mostly worn off,
and never hairy. Whorls 5¼ to 5¾, the last
whorl barely contracted behind the apertural lip. Aperture lacks a parietal denticle.
Apertural lip white, strongly flared and
basally thickened, with a slight bulging callus. Genitalia: Pilsbry 1940.

5 mm

   Some individuals in this species, like in
Allogona townsendiana, have a paler, straw-
yellow shell. *A. ptychophora* is slightly
smaller than *A. townsendiana* and its shell
sculpture is usually less dimpled and more
distinctly rib-like.

Distribution
Southern British Columbia south through Montana, Idaho, and west down the Columbia River valley of Washington and Oregon. *Allogona ptychophora* is common in B.C., and generally spread throughout the southern Kootenay, Columbia and Elk river drainages, at least as far north as Revelstoke; there are also museum records from Vernon and Salmon Arm.

Natural History
These snails are common, living under leaf litter, grass, rocks and logs, and on ground in the open or in woods. During winter hibernation, most position their shells with the aperture facing upward – this orientation appears to be important for survival.

Etymology
Allogona: "different genitalia"; *Dysmedoma*: "sunset-house", in reference to the western distribution of this subgenus; *ptychophora*: "bearing folds".

Selected References
Carney 1966, Pilsbry 1940.

Allogona (*Dysmedoma*) *townsendiana*   Oregon Forestsnail (I. Lea, 1839)
Synonyms: *Helix townsendiana* I. Lea, 1839; *Polygyra townsendiana brunnea* Vanatta, 1924; *Allogona townsendiana frustrationensis* Pilsbry, 1940.

Description
Shell 23–30 mm wide, heliciform and pale brown or straw-yellow, with coarse, irregular, lighter-coloured wrinkle-like axial riblets, exceedingly fine, wavy spiral striae and, often, dimpled sculpture. Periostracum sometimes mostly worn off, revealing the chalky shell underneath, and never hairy. Whorls 5¼ to 6, the last whorl barely contracted behind the apertural lip. No parietal denticle. Apertural lip white, thickened and flared, with a slightly bulging basal callus. Colour photograph C-29. Genitalia: Pilsbry 1940.

Similar to *Allogona ptychophora*, which is smaller, has more neatly placed axial riblets and seldom has strongly developed dimpled sculpture.

Distribution
Southwestern British Columbia to northwestern Oregon. In B.C., *Allogona townsendiana* occurs in a small area along the Chilliwack and Fraser river valleys from near Hope to Mission and Langley. There are also a few records from southern Vancouver Island.

Natural History
This species lives in moist, lowland deciduous woods with dense undergrowth, often in patches of Stinging Nettle. Adults are most likely seen in late spring and early summer when they are most active. During this time, they lay eggs in clusters in shallow depressions in the soil. Young snails are elusive and rarely seen.

Etymology
Allogona: "different genitalia"; *Dysmedoma*: "sunset-house", in reference to the western distribution of this subgenus; *townsendiana*: named for John K. Townsend (1809–51), Philadelphia ornithologist and naturalist, who in 1834 accompanied Thomas Nuttall west to Oregon, making collections.

5 mm

Remarks
In 2002 *Allogona townsendiana* was placed on the Canadian list of Species at Risk as "Endangered". Although in the American portion of its range this species is reasonably secure, in Canada it is restricted to relatively small, isolated habitats on agricultural and suburban lands. Threats include continued deforestation and habitat modification due to urban expansion.

Selected References
Branson 1977, Cameron 1986, Pilsbry 1940.

Key to species of *Cryptomastix*

1a   Shell width less than 9 mm, lip strongly flared but not recurved; parietal denticle long and curved (in basal view) .... *Cryptomastix germana*

1b   Shell width less than 9 mm, lip flared and recurved; parietal denticle not long and curved (in basal view) 2

2a   Shell width up to 26 mm; historically on southern Vancouver Island and adjacent mainland; heliciform and not particularly flattened .... *C. devia*

2b   Shell width up to 19 mm; southern interior distribution; heliciform to depressed-heliciform .... *C. mullani*

Cryptomastix (*Cryptomastix*) *devia*   **Puget Oregonian** (Gould, 1846)

Synonym: *Helix devia* Gould, 1846; *H. baskervillei* Pfeiffer, 1850.

Description

Shell 20–23 mm wide, heliciform and pale brown, nearly smooth, with a few irregular, very low axial riblets (strongest next to the suture), and fine incremental and spiral striae. Periostracum not hairy in adults. Spire moderately elevated. About 6 whorls. Aperture has a prominent white parietal denticle, not curved but peg-like in basal view. Apertural lip white, broadly expanded and recurved, with a ridge-like swelling at the base. Animal light brown, sometimes tinged with violet. Genitalia: Pilsbry 1940.

This species is most similar to *Cryptomastix mullani*, which is smaller and flatter.

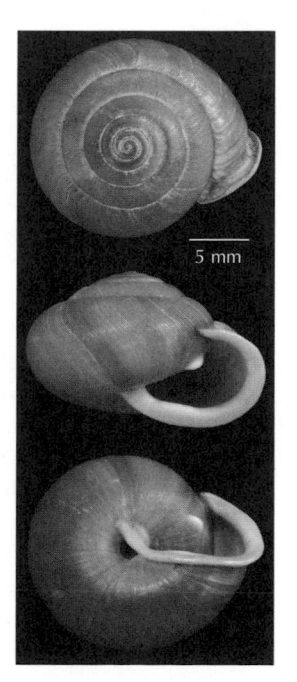

Distribution

Southwestern British Columbia to the Columbia River Gorge, in Washington and Oregon. This species is known from British Columbia by only three old collections: Vancouver Island (no specific locality),

Esquimalt and Sumas Prairie. *Cryptomastix devia* has a patchy distribution throughout its range.

Natural History
This species lives in moist forested sites at low elevations.

Etymology
Cryptomastix: "hidden flagellum" – these were anatomically distinguished from the eastern and central North American genus *Triodopsis* by the presence of a vestigial structure that Henry Pilsbry called a *flagellum*; *devia*: "out of the way" or "solitary".

Remarks
As there are no recent records of this species from British Columbia since the late 1800s, it is believed that the species no longer occurs in Canada and in 2002 was placed on the Canadian list of Species at Risk as "Extirpated".

Selected References
Branson 1980, Dall 1905, Pfeiffer 1850, Pilsbry 1940, Taylor 1889, Vagvolgyi 1968.

Cryptomastix (Cryptomastix)   Coeur d'Alene Oregonian *mullani* **(Bland & J.G. Cooper, 1861)**
Synonyms: *Helix mullani* Bland & J.G. Cooper, 1861; ?*Polygyra mullani olneyae* Pilsbry, 1891; see Remarks.

Description
Shell 12–17 mm wide, heliciform to depressed-heliciform, brown and nearly smooth, sometimes with fine incremental and weak spiral striae. Periostracum not hairy in adults. About 16 whorls. Parietal denticle low to strongly developed, but peg-like and not long and curved when in basal view. Apertural lip thickened, strongly recurved and with obscurely denticle-like thickenings. Genitalia: Pilsbry 1940.

Similar to *Cryptomastix devia*, which has a larger, less flattened shell. *C. germana* is smaller, with a flared but not recurved apertural lip and a long, curved parietal denticle.

Distribution
Southeastern British Columbia to Oregon, and east to western Montana.

Natural History
This species lives under rocks and logs and on forest vegetation.

Etymology
Cryptomastix: "hidden flagellum" – see Etymology under *C. devia*; *mullani*: after Captain John Mullan, U.S. Army engineer, who surveyed a road over the Bitterroot Mountains in 1853–56.

Remarks
The species name *Cryptomastix mullani* is used here in the broad sense, because there are several named forms or subspecies that require anatomical investigation. Neither the classifications of J. Vagvolgyi (1968) nor Henry Pilsbry (1940) provide an adequate treatment of this genus, and at least some of the named forms or subspecies of *C. mullani* are potentially valid. The taxon, *C. mullani olneyae* (Pilsbry, 1891) is the only subspecies reported from British Columbia at this time.

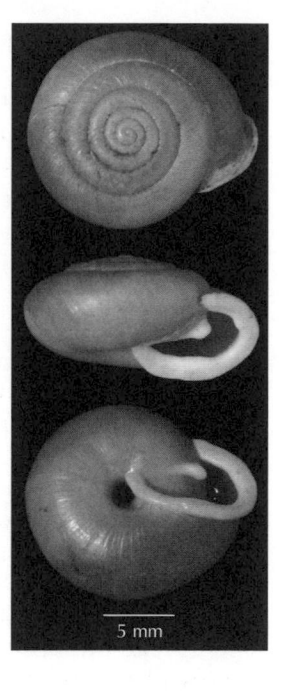

5 mm

Selected References
Pilsbry 1940, Vagvolgyi 1968.

Cryptomastix (Micranepsia) germana   **Pygmy Oregonian**
(Gould *in* A. Binney, 1851)
Synonyms: *Helix germana* Gould *in* A. Binney, 1851; *Polygyra germana vancouverinsulae* Pilsbry & Cooke, 1922.

Description
Shell 6.5–8.3 mm wide, heliciform and light to dark brown, with fine incremental wrinkles and striae; periostracum hairy, usually persisting in adults; 4 to 5 whorls. Parietal denticle well developed, long and curved in basal view. Apertural lip white, strongly flared

but not thickened or recurved. Umbilicus tiny and mostly concealed by the columellar lip. Animal dark greyish brown. Colour photograph on front cover. Genitalia: Pilsbry 1940.

Juveniles of this species and *Vespericola columbianus* may be confused, but the whorls are more tightly coiled and the periostracal hairs thicker and more sparse in *Cryptomastix germana.* This is the smallest of the *Cryptomastix* species, which also has a unique anatomy and a flared but never thickened or recurved apertural lip.

Distribution
Moresby Island, British Columbia, to southwest Oregon. It occurs west of the Cascade and Coast ranges in B.C.

5 mm

Natural History
This species is common, but seldom abundant, in moist deciduous or mixed coastal forests, where it lives under logs and Sword Ferns and in leaf litter. It is a lowland species but has been recorded at 1372 metres elevation in the Olympic Mountains, Washington.

Etymology
Cryptomastix: "hidden flagellum" (see Etymology under *C. devia)*; *micranepsia*: "little cousin"; *germana*: "sister" – Gould thought this species to be closely related to an eastern American snail.

Selected References
Branson 1977, 1980; Cameron 1986; Forsyth 2000; Pilsbry and Cooke 1922; Vagvolgyi 1968.

Vespericola columbianus   Northwest Hesperian
(I. Lea, 1839)

Synonyms: *Helix columbiana* I. Lea, 1839; *Polygyra columbiana pilosa* of authors, not Henderson, 1928; *Vespericola columbiana latilabrum* Pilsbry, 1940.

Description

Shell 10–17 mm wide, heliciform, light or dark brown and densely covered in fine, erect hairs. Aperture lacks a parietal denticle; apertural lip white or pale brown, strongly expanded and recurved. Umbilicus small and partially concealed by the apertural lip. Animal pale brown; mantle with dark blotches visible through the shell, especially in juveniles. Colour photograph C-4. Genitalia: Roth and Miller 1993.

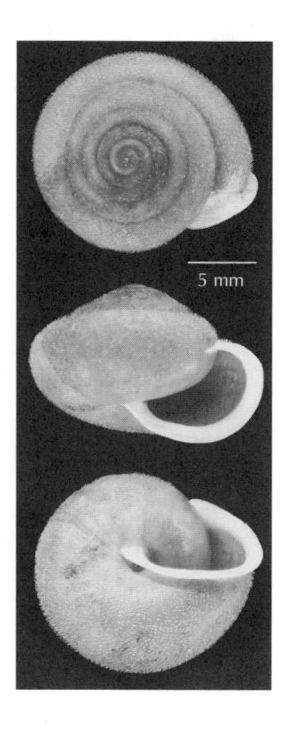

The size of the shell and elevation of the spire vary within and between populations. The hairy shell accumulates dirt and humus, often making living snails difficult to spot. Sometimes the hairs wear off, leaving only microscopic hair scars (colour photograph C-5). *Cryptomastix germana* also has a hairy shell, but adults are smaller and have a strong parietal denticle; its juveniles have a tighter coiling shell and more scattered hairs.

Distribution

Unalaska Island, Alaska to Oregon (doubtfully to northwestern California). This species is widespread and common throughout British Columbia west of the Coast and Cascade ranges, extending further inland up major river valleys.

Natural History

Vespericola columbianus lives in coniferous, deciduous and mixed forests and at open sites, where it finds shelter under vegetation, wood and rocks. It is often abundant under driftwood and logs adjacent to the seashore, but it also occurs at higher elevations.

Etymology
Vespericola: "west-dwelling"; *columbianus*: of the Columbia River.

Remarks
Until recently, most British Columbia, Washington and Alaska *Vespericola* were attributed to *V. columbianus pilosus* (Henderson, 1928), but Barry Roth and Walter Miller showed that *V. pilosus*, which occurs only near San Francisco region of central California, is a separate species. *Vespericola* species are primarily recognized from one other by their reproductive anatomy.

Selected References
Forsyth 2000d, Pilsbry 1940, Roth and Miller 1993.

Unconfirmed species of Polygyridae

Allogona lombardii **Smith, 1943**   **Selway Forestsnail**
A.G. Smith (1943) recorded this species at Boswell, on the east shore Kootenay Lake, but this is probably erroneous and *A. lombardii* likely does not occur in British Columbia.

Family Thysanophoridae

Microphysula cookei **(Pilsbry, 1922)**   **Vancouver Snail**
Synonym: *Zonitoides cookei* Pilsbry, 1922.

Description
Shell 4.4 mm wide, flattened-heliciform, translucent whitish, with a nearly flat spire. About 5 whorls, tightly and almost flatly coiled. Protoconch smooth; subsequent whorls marked with weak, microscopic incremental striae and fine, close-spaced spiral striae. Umbilicus about a quarter the width of the shell. Genitalia: unknown.

This species so similar to *Microphysula ingersollii* that they may be difficult to separate. *M. cookei* is slightly smaller and its whorls are not as tightly coiled.

Distribution
Anchorage, Alaska, to the Skeena, Boundary, Coast and Vancouver Island ranges, British Columbia, south to the Coast and Cascade mountains of Washington.

Natural History
This species lives in wet montane coniferous forests and in vegetated rockslides at and below the tree line.

Etymology
Microphysula: "little bubble"; *cookei*: in honour of Dr Charles Montague Cooke, Jr (1874–1948) of Hawaii, a friend and associate of Henry Pilsbry. In July and August 1918, Cooke collected land snails at several localities on central Vancouver Island; *M. cookei* was named from specimens that he collected at Cameron Lake.

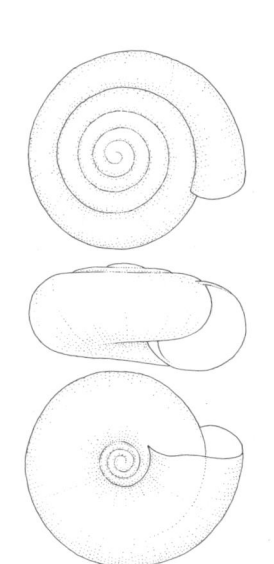

1 mm

Selected References
Forsyth 2001d, Pilsbry 1940.

Microphysula ingersollii (Bland, 1875) Spruce Snail
Synonyms: *Helix ingersollii* Bland, 1875; *Thysanophora ingersolli convexior* Ancey, 1887; *T. ingersolli meridionalis* Pilsbry & Ferriss, 1910.

Description
Shell 4.8 mm wide, flattened-heliciform, translucent whitish, with a nearly flat spire. About 5 tightly coiled whorls. Protoconch smooth; subsequent whorls marked with weak, microscopic incremental striae and fine, close-spaced spiral striae. Umbilicus about a quarter the width of the shell. Genitalia: Pilsbry 1940.

The shells of this species and *Microphysula cookei* are nearly identical. *M. ingersollii* is slightly larger and its whorls are more tightly coiled.

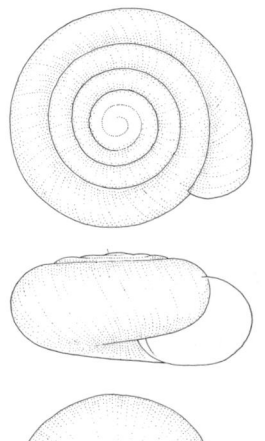

Distribution
Southeast British Columbia and southwest Alberta to New Mexico and Arizona, and east to northwest Wyoming. In B.C., this species is common in the mountains of the Kootenays, but has been found north to Pine Pass, Hart Ranges.

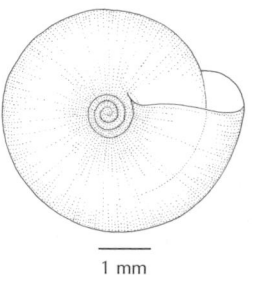

Natural History
This species lives in subalpine meadows, rockslides, spruce forests and Trembling Aspen groves where it occurs in moist leaf litter and under dead wood and rocks. It is particularly abundant in wet, montane localities with limestone.

1 mm

Etymology
Microphysula: "little bubble"; *ingersollii*: after Ernest Ingersoll, a U.S. Geological Survey naturalist, who discovered the species.

Selected References
Bequaert and Miller 1973, Pilsbry 1940.

Family Bradybaenidae

Monadenia (Monadenia) fidelis (J. E. Gray, 1834)

Pacific Sideband

Synonyms: *Helix fidelis* J.E. Gray, 1834; *H. nuttalliana* I. Lea, 1839; *H. oregonensis* I. Lea, 1839; *Monadenia fidelis columbiana* Pilsbry, 1939; *M. fidelis semialba* Henderson, 1929.

Description

Shell 22–36 mm wide, heliciform, usually chestnut brown with a narrow, pale yellow band at the periphery, a narrow, dark brown band above this, and a dark brown base. Whorls 5¼ to 6¾, the last whorl descending before the aperture. Aperture whitish inside with darker pigment showing through. Apertural lip is slightly thickened, and pale brown or purplish brown. Umbilicus small. Animal rosy brown, with greyish spaces between the tubercles; behind the head are some sparse, black reticulations. Mantle has a bold rust-coloured band. Sole of the foot is pale grey. Colour photograph C-2. Genitalia: Pilsbry 1939.

5 mm

There is considerable variation in the colour and banding of this species. The yellow band, for example, varies in width, and in some populations, it may even extend below the periphery and onto the underside of the shell. Some individuals may be pale yellow without substantially darker banding (bottom shell in photograph). The introduced snail, *Cepaea nemoralis*, is smaller and lacks an umbilicus in adults.

Distribution

Sitka, Alaska, to Cape Mendocino, California; west of the Coast and Cascade mountains in British Columbia.

Natural History
Monadenia fidelis lives in deciduous, coniferous or mixed forests but also in open woods and grassy areas. It has even been found under low mats of stunted Garry Oak on a small, rocky islet in the Gulf Islands. *M. fidelis* is most often encountered in late spring when adult snails are crawling on the ground in the open or climbing the trunks of shrubs or trees – snails have been observed in trees as high as 6.7 metres above the ground.

Etymology
Monadenia: "one gland", a reference to the presence of a single mucus gland; *fidelis:* "dependable" or "faithful".

Remarks
Malacologists recognize several subspecies of *Monadenia fidelis*, but none other than the typical subspecies, *M. fidelis fidelis*, has been recorded in British Columbia.

Selected References
Branson 1984, Dall 1905, Jackson 1923, Pilsbry 1939, Roth 1981, Roth and Sadeghian 2003, Webb 1952.

Family Helicidae

***Cepaea* (*Cepaea*) *nemoralis* (Linnaeus, 1758)**   **Grovesnail**
Synonym: *Helix nemoralis* Linnaeus, 1758.

Description
Shell 20–25 mm wide, heliciform, opaque and somewhat shiny; brown, orange or yellow and usually with 1–5 blackish or dark brown spiral bands, which may be fused together or entirely absent. Whorls 4½ to 5¼, the last whorl descending before the aperture. Aperture wider than its height. Apertural lip purplish brown, thickened and slightly recurved. Juveniles have a narrow umbilicus and adults have no umbilicus. Animal cream or pale brown, but darker on the head and tentacles. Colour photograph C-3. Genitalia: Adam 1960, Gittenberger et al. 1984.

The shell of *Cepaea nemoralis* varies remarkably in colour. It can have zero to five bands, though three is most common. Compare this species with other large land snails, especially *Cornu aspersum* and *Monadenia fidelis*.

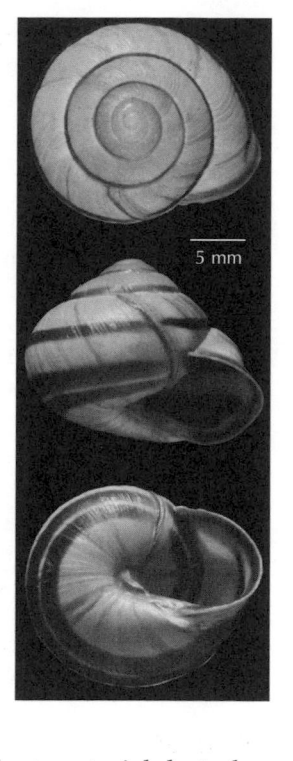

5 mm

Distribution
Native to central and western Europe and widely introduced to North America. In British Columbia, this species is locally common in Vancouver, the Fraser Valley, the Gulf Islands, southern Vancouver Island, the Okanagan Valley and Nelson. In Washington, it occurs around the Puget Sound region.

Natural History
Cepaea nemoralis is one of the more familiar snails in urban areas where it is widespread, in gardens and vacant lots. It is often seen crawling or aestivating on the trunks of trees, sometimes well off the ground. It primarily eats dying or dead plant material, but also green plants, dead animal remains, fungi, mosses and possibly live aphids and thrips. In Britain, this species takes three years to reach adult size.

The colour patterns of these shells are genetically determined and there is a huge body of literature on the selective and non-selective influences on this polymorphism.

Etymology
Cepaea: "an onion"; *nemoralis*: "from a grove" or "woodland".

Selected References
Davison 2000, Draycot 1961, Forsyth 1999b, Jones et al. 1977, Kerney 1999, Kerney and Cameron 1979, Hanna 1966, Honek 1995, Pilsbry 1939, Richardson 1975, Singh 1981, Williamson 1979, Williamson and Cameron 1976, Wolda 1970, Wolda et al. 1971; there is much more literature on this species.

Cornu aspersum (Müller, 1774)   Brown Gardensnail
Synonyms: *Helix aspersa* Müller, 1774.

Description
Shell 27–32 mm wide, heliciform and yellowish brown with darker brown spiral bands interrupted by irregular lighter markings and axial streaks, with incremental striae, fine raised irregular axial riblets, irregular dimpling, and very fine spiral striae. About 3¾ to 4¼ whorls. Aperture large and rounded. Apertural lip white, a little thickened and slightly recurved. Umbilicus usually absent in adults or infrequently open as a narrow slit. Animal dark grey to pale brown-ochre with tubercles yellowish; mantle blackish, speckled with greyish yellow. Genitalia: Barker 1999, Gittenberger et al. 1984.

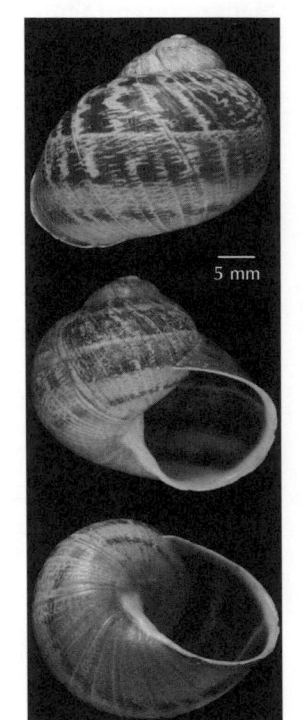

5 mm

    This species is larger than *Cepaea nemoralis*, its shell has fewer whorls, its aperture is more rounded and its colour pattern is altogether different.

Distribution
Native of southern Europe, and introduced to North America, South America, Africa, East Asia, Oceania, New Zealand and Australia. Sporadic populations occur in Victoria and elsewhere on southern Vancouver Island, the Gulf Islands and Greater Vancouver, as well as around Puget Sound in Washington State.

Natural History
In British Columbia, *Cornu aspersum* lives in gardens, along roadsides and in vacant lots. These snails feed on both live and dead plants.

Etymology
Cornu: "a horn"; *aspersum*: "scattered" or "sprinkled", likely describing the speckled and streaked colour pattern on the shell.

Remarks

Cornu aspersum is one of several large European snails that humans eat as escargot.

Selected References

Barker 1999, Bequaert and Miller 1973, Chung 1987, Desbuquois and Madec 1998, Ellis 1969, Forsyth 1999c, Guiller et al. 2001, Hanna 1966, Herzberg and Herzberg 1962, Kerney 1999, Kerney and Cameron 1979, Landolfa 2002, Pilsbry 1939, Robinson 1999, Roth and Sadeghian 2003; there is much more literature on this species.

REFERENCES AND BIBLIOGRAPHY

Adam, W. 1960. *Mollusques. Tome 1. Mollusques terrestres et dulcicoles.* Bruxelles:
Institut Royal des Sciences Naturelles de Belgique.

Anderson, J., and R. Mehl. 1992. The landsnail *Euconulus fulvus* (Gastropoda:
Pulmonata) found on Willow Grouse (*Lagopus lagopus*). *Fauna Norwegica*
Series A, 13:37–38.

Arad, Z., S. Goldenberg, and J. Heller. 1998. Short- and long-term resistance to
desiccation in a minute litter-dwelling land snail *Lauria cylindracea*
(Pulmonata: Pupillidae). *Journal of Zoology (London)* 246:75–81.

Armbruster, G. 1994. The taxonomically relevant parts of the male genitalia of
Cochlicopa (Gastropoda: Pulmonata: Cochlicopidae): seasonal variability
within two field populations and observations under laboratory conditions.
Malakologische Abhandlungen Staatliches Museum für Tierkunde Dresden 17:47–56.

———. 1995. Univariate and multivariate analyses of shell variables within
the genus *Cochlicopa* (Gastropoda: Pulmonata: Cochlicopidae). *The Journal of
Molluscan Studies* 61:225–235.

———. 1997. Evaluations of RAPD markers and allozyme patterns: evidence
for morphological convergence in the morphotype of *Cochlicopa lubricella*
(Gastropoda: Pulmonata: Cochlicopidae). *The Journal of Molluscan Studies*
63:379–388.

———. 2001. Selection and habitat-specific allozyme variation in the self-fertilizing land snail *Cochlicopa lubrica* (O. F. Müller). *Journal of Natural History*
35:185–199.

Armbruster, G.F.J., and Bernhard, D. 2000. Taxonomic Significance of
Ribosomal ITS-1 sequence markers in self-fertilizing land snails of
Cochlicopa (Stylommatophora, Cochlicopidae). *Mitteilungen aus dem Museum
für Naturkunde in Berlin, Zoologische Reihe* 76:11–18.

Armbruster, G., and M. Schlegel. 1994. The landsnail species of *Cochlicopa*
(Gastropoda: Pulmonata): presentation of taxon-specific allozyme patterns,
and evidence for a high level of self-fertilization. *Journal of Zoological
Systematics and Evolutionary Research* 32:282–296.

Backeljau, T. 1985. Estimation of genic similarity within and between *Arion hortensis sensu lato* and *Arion intermedius* by means of isoelectric focused esterase
patterns in hepatopancreas homogenates (Mollusca, Pulmonata: Arionidae).
Zeitschrift für zoologische Systematik und Evolutionsforschung 23:38–49.

Backeljau, T., and M. Van Beeck. 1986. Epiphallus anatomy in the *Arion hortensis* species aggregate (Mollusca, Pulmonata). *Zoologica Scripta* 15:61–68.

Backeljau, T., L. De Bruyn, H. De Wolf, K. Jordaens, S. Van Dongen, and B. Winnepenninckx. 1997. Allozyme diversity in slugs of the *Carinarion* complex (Mollusca, Pulmonata). *Heredity* 78:445–451.

Backeljau, T., K. Jordaens, H. DeWolf, T. Rodríguez, and B. Winnepenninckx. 1997. Albino-like *Carinarion* identified by protein electrophoresis (Pulmonata: Arionidae). *The Journal of Molluscan Studies* 63:559–563.

Backeljau, T., B. Winnepenninckx, K. Jordaens, H. DeWolf, K. Breugelmans, C. Parejo, and T. Rodriguez. 1996. Protein electrophoresis in arionid taxonomy. *British Crop Protection Council Symposium Proceedings* 66:21–28.

Backhuys, W. 1975. *Zoogeography and taxonomy of the land and freshwater Molluscs of the Azores.* Backhuys and Meesters, Amsterdam.

Baird, W. 1863. Descriptions of some new species collected at Vancouver Island and in British Columbia by J.K. Lord. *Proceedings of the Zoological Society of London for 1863,* pp. 66–71.

Baker, F.C. 1902. The Mollusca of the Chicago area. *The Chicago Academy of Sciences, Natural History Survey, Bulletin* 3:131–410, pls. 28–36.

Baker, H.B. 1930a. New and problematic West American land-snails. *The Nautilus* 43:95–101, 121–128.

———. 1930b. The North American Retinellae. *Proceedings of the Academy of Natural Sciences of Philadelphia* 83:193–219.

———. 1930c. The land snail genus *Haplotrema. Proceedings of the Academy of Natural Sciences of Philadelphia* 82:405–425, pls. 33–35.

———. 1931. Nearctic vitreine land snails. *Proceedings of the Academy of Natural Sciences of Philadelphia* 83:85–117.

———. 1941. Zonitid snails from Pacific Islands, parts 3 and 4. 3. Genera other than Microcystinae. 4. Distribution and indexes. *Bernice P. Bishop Museum, Bulletin* 166:203–370, pls. 43–65.

Bank, R.A., and E. Gittenberger. 1985. Notes on Azorean and European *Carychium* species (Gastropoda, Basommatophora, Ellobiidae). *Basteria* 49:85–99.

Barker, G.M. 1999. Naturalised terrestrial Stylommatophora (Mollusca: Gastropoda). *Fauna of New Zealand / Ko te Aitanga Pepeke o Aotearoa* 38.

———. 2001. *The Biology of Terrestrial Molluscs.* Wallingford, U.K.: CABI Publishing.

———. 2002. *Molluscs as crop pests.* Wallingford, U.K.: CABI Publishing.

Baur, A. 1987. The minute land snail *Punctum pygmaeum* (Draparnaud) can reproduce in the absence of a mate. *The Journal of Molluscan Studies* 53:112–113.

Baur, B., S. Ledergerber and H. Kothbauer. 1997. Passive dispersal on mountain slopes: shell shape related differences in downhill rolling in the land snails *Arianta arbustorum* and *Arianta chamaeleon* (Helicidae). *The Veliger* 40:84–85.

Bayne, C.J. 1973. Physiology of the pulmonate reproductive tract: location of spermatozoa in isolated, self-fertilizing succineid snails (with a discussion of pulmonate tract terminology). *The Veliger* 16:169–175.

Beetle, D.A. 1957. The Mollusca of Teton County, Wyo. *The Nautilus* 71:12–21.

Bequaert, J.C., and W.B. Miller. 1973. *The mollusks of the arid Southwest: with an Arizona check list.* Tucson: University of Arizona Press.

Berry, S.S. 1919. Mollusca of Glacier National Park, Montana. *Proceedings of the Academy of Natural Sciences of Philadelphia* 1919: 195–205, pls 9–10.

———. 1922. Land snails from the Canadian Rockies. *Victoria Memorial Museum, Bulletin* 36.

Blood, D.A. 1963. Parasites from California bighorn sheep in southern British Columbia. *Canadian Journal of Zoology* 41:913–918.

Boag, D.A. 1985. Microdistribution of three genera of small terrestrial snails (Stylommatophora: Pulmonata). *Canadian Journal of Zoology* 63:1089–1095.

Boag, D.A., and W.D. Wishart. 1982. Distribution and abundance of terrestrial gastropods on a winter range of bighorn sheep in southwestern Alberta. *Canadian Journal of Zoology* 60:2633–2640.

Borreda, V., M.A. Collado, J. Blasco, and J.S. Espin. 1996. Slugs (Gastropoda, Pulmonata) of Andorra. *Iberus* 14:25–36.

Branson, B.A. 1969. Distribution notes on western and southern snails. *Sterkiana* 36:21.

———. 1972. *Hemphillia dromedarius*, a new arionid slug from Washington. *The Nautilus* 85:100–106.

———. 1975. *Radiodiscus hubrichti* (Pulmonata: Endodontidae) new species from the Olympic Peninsula, Washington. *The Nautilus* 89:47–48.

———. 1977. Freshwater and terrestrial Mollusca of the Olympic Peninsula, Washington. *The Veliger* 19:310–330.

———. 1980. Collections of gastropods from the Cascade Mountains of Washington. *The Veliger* 23:171–176.

Branson, B.A., and R.M. Branson 1984. Distributional records for terrestrial and freshwater Mollusca of the Cascade and Coast ranges, Oregon. *The Veliger* 26:248–257.

Branson, R.M. 1983. Geographical variation of banding and color morphs in *Monadenia fidelis* (Gray 1834). *The Veliger* 25:349–355.

Bulman, K. 1990. Life history of *Carychium tridentatum* (Risso, 1826) (Gastropoda: Pulmonata: Ellobiidae) in the laboratory. *The Journal of Conchology* 33:321–333.

Brunson, R.B., and N. Kevern. 1964. Observations on a colony of *Magnipelta*. *The Nautilus* 77:23–27.

Brunson, R.B., and U. Osher. 1957. *Haplotrema* from western Montana. *The Nautilus* 70:121–123.

Burch, J.B. 1962. *How to know the eastern land snails: pictured keys for determining the land snails of the United States occurring east of the Rocky Mountain Divide.* Dubuque, Iowa: William C. Brown Company.

Caesar, N.H. 1946. A roundup of *Cochlicopa lubrica. The Nautilus* 60:72.

Cain, A.J., and M.H. Williamson. 1958. Variation and specific limits in the *Arion ater* aggregate. *Proceedings of the Malacological Society of London* 33:72–86.

Cameron, R.A.D. 1982. Life histories, density and biomass in a woodland snail community. *The Journal of Molluscan Studies* 48:159–166.

———. 1986. Environment and diversities of forest snail faunas from coastal British Columbia. *Malacologia* 27:341–355.

Cameron, R.A.D., B. Eversham, and N. Jackson. 1983. A field key to the slugs of the British Isles. *Field Studies* 5:807–824.

Campbell, R.W., and D. Stirling. 1968. Notes on the natural history of Cleland Island, British Columbia, with emphasis on the breeding bird fauna. *Report of the Provincial Museum, 1967*: HH25–HH43.

Carl, G.C., and C.J. Guiguet. 1972. *Alien animals in British Columbia*, 2nd edition, revised by C.J. Guiguet. British Columbia Provincial Museum Handbook 14. Victoria: British Columbia Provincial Museum.

Carney, W.P. 1966. Mortality and apertural orientation in *Allogona ptychophora* during winter hibernation in Montana. *The Nautilus* 79:134–136.

Chelazzi, G., G. Le Voci, and D. Parpagnoli. 1988. Relative importance of airborne odours and trails in the group homing of *Limacus flavus* (Linnaeus) (Gastropoda, Pulmonata). *The Journal of Molluscan Studies* 54:173–180.

Chung, D.J.D. 1987. Courtship and dart shooting behaviour of the land snail *Helix aspersa*. *The Veliger* 31:24–39.

Colville, B., L. Lloyd-Evans, and A. Norris. 1974. *Boettgerilla pallens* Simroth, a new British species. *The Journal of Conchology* 28:203–208.

Dall, W.H. 1905. Land and fresh water mollusks. *Harriman Alaska Expedition* 13:1–171, pls. 1–2.

Davies, S.M. 1977. The *Arion hortensis* complex, with notes on *A. intermedius* Normand (Pulmonata: Arionidae). *The Journal of Conchology* 29:173–187.

———. 1979. Segregates of the *Arion hortensis* complex (Pulmonata: Arionidae), with the description of a new species, *Arion owenii*. *The Journal of Conchology* 30:123–127.

Davison, A. 2000. The inheritance of divergent mitochondria in the land snail, *Cepaea nemoralis*. *The Journal of Molluscan Studies* 66: 143–147.

Denny, M.W. 1980. Locomotion: the cost of gastropod crawling. *Science* 208:1288–1290.

Desbuquois, C., and L. Madec. 1998. Within-clutch egg cannibalism variability in hatchlings of the land snail *Helix aspersa* (Pulmonata: Stylommatophora): influence of two proximate factors. *Malacologia* 39:167–173.

Deyrup-Olsen, I., A.W. Martin, and R.T. Paine. 1986. The autotomy escape response of the terrestrial slug *Prophysaon foliolatum* (Pulmonata: Arionidae). *Malacologia* 227:307–311.

Dodds, C.J., I.F. Henderson and P. Watson. 1997. Induction of activity in the olfactory nerve of the slug *Deroceras reticulatum* (Müller) in response to volatiles emitted by carabid beetles. *The Journal of Molluscan Studies* 63:297–98.

Drake, R.J. 1963. The history of nonmarine malacology in British Columbia. *National Museum of Canada, Bulletin* 185.

Draycot, W.M. 1961. Mollusks introduced into British Columbia. *The Canadian Field-Naturalist* 75:164.

Dundee, D.S., P.H. Phillips, and J.D. Newsom. 1967. Snails on migratory birds. *The Nautilus* 80:89–91.

Dutra-Clarke, A.V.C., C. Williams, R. Dickstein, N. Kaufer, and J.R. Spotila. 2001. Inferences on the phylogenetic relationships of Succineidae (Mollusca, Pulmonata) based on 18S rRNA gene. *Malacologia* 43:223–236.

Ellis, A.E. 1969. *British Snails. A guide to the non-marine Gastropoda of Great Britain and Ireland, Pleistocene to Recent*. Rev. ed. Oxford University Press.

Emberton, K.C. 1995. When shells do not tell: 145 million years of evolution in North America's polygyrid land snails, with a vision and conservation priorities. *Malacologia* 37:69–110.

Emberton, K.C., G.S. Kunico, G.M. Davis, S.M. Philips, K.M. Monderewicz, and Y.H. Guo. 1990. Comparison of recent classifications of stylommatophoran land-snail families and evaluation of large-ribosomal-RNA sequencing for their phylogenetics. *Malacologia* 31:327–352.

Falkner, G., T.E.J. Ripken, and M. Falkner. 2002. Mollusques continentaux de France. Liste de Référence annotée et Bibliographie. *Patrimoines naturels* 52.

Foltz, D.W., B.M. Schaitkin, and R.K. Selander. 1982. Gametic disequilibrium in the self-fertilizing slug *Deroceras laeve*. *Evolution* 36:80–85.

Forcart, L. 1955. Die nordischen Arten der Gattung *Vitrina*. *Archiv für Molluskenkunde* 84:155–166.

———. 1959a. Die palaeartischen Arten des Genus *Columella*. *Verhandlungen der Naturforschenden Gesellschaft in Basel* 70:7–18.

———. 1959b. Taxionomische Revision paläarktischer Zonitinae, II. Anatomisch untersuchte Arten des Genus *Aegopinella* Lindholm. *Archiv für Molluskenkunde* 88:7–33.

Forsyth, R.G. 1999a. Lindeman Lake, British Columbia, type locality of *Zonitoides randolphi* Pilsbry. *The Veliger* 42:286.

———. 1999b. Distributions of nine new or little-known exotic land snails in British Columbia. *The Canadian Field-Naturalist* 113:559–568.

———. 2000. The land snail *Cryptomastix germana* (Gastropoda: Polygyridae) in the Queen Charlotte Islands, British Columbia: a range extension north from Vancouver Island. *The Canadian Field-Naturalist* 114:316–317.

———. 2001a. Re-identification of slugs from seabird nesting burrows off the west coast of Vancouver Island. *The Festivus* 33:9–10.

———. 2001b. A note on the distribution of *Striatura pugetensis* in British Columbia. *The Festivus* 33:57–58.

———. 2001c. First records of the European land slug *Lehmannia* in British Columbia, Canada. *The Festivus* 33:75–78.

———. 2001d. New records of land snails from the mountains of northwestern British Columbia. *The Canadian Field-Naturalist* 115:223–228.

———. 2004. *Gastrocopta* in British Columbia (Mollusca: Pulmonata: Vertiginidae). *The Festivus* 36:53-55.

Forsyth, R.G., J.M.C. Hutchinson, and H. Reise. 2001. *Aegopinella nitidula* (Draparnaud, 1805) (Gastropoda: Zonitidae) in British Columbia – first confirmed North American record. *American Malacological Bulletin* 16:65–69.

Franzen, D.S. 1982. *Succinea avara* Say from the Southern Great Plains of the United States. *The Nautilus* 96:82–88.

———. 1985. Anatomy of *Oxyloma nuttalliana chasmodes* Pilsbry. *The Nautilus* 99:134–139.

Frest, T.J., and E.J. Johannes. 2001. An annotated checklist of Idaho land and freshwater mollusks. *Journal of the Idaho Academy of Science* 36:1–51.

Frest, T.J., and R.S. Rhodes, II. 1982. *Oxychilus draparnaldi* [sic] in Iowa. *The Nautilus* 96:36–39.

Garrido, C., J. Castillejo, and J. Iglesias. 1995a. The *Arion subfuscus* complex in the eastern part of the Iberian Peninsula, with redescription of *Arion subfuscus* (Draparnaud 1805). *Archiv für Molluskenkunde* 124:108–118.

———. 1995b. The spermatophore of *Arion intermedius* (Pulmonata: Arionidae). *The Journal of Molluscan Studies* 61:127–133.

Gelperin, A. 1974. Olfactory basis of homing behaviours in the giant garden slug, *Limax maximus*. *Proceedings of the National Academy of Sciences of the U.S.A.* 71:966–970.

Gerber, J. 1996. Revision der Gattung *Vallonia* Risso 1826 (Mollusca: Gastropoda: Valloniidae). *Schriften zur Malakozoologie aus dem Haus der Natur-Cismar* 8:1–227.

Getz, L.L. 1959. Notes on the ecology of slugs: *Arion circumscriptus, Deroceras reticulatum*, and *D. laeve. The American Midland Naturalist* 61:485–498.

Gittenberger, E., W. Backhuys, and T.E.J. Ripken. 1984. *De Landslakken van Nederland*. Koninklijke Nederlandse Natuurhistorische Vereniging, Amsterdam.

Giusti, F., and G. Manganelli. 1990. Notulae malacologicae, XLIV. A neotype for *Agriolimax caruanae* Pollonera 1891 (Pulmonata: Agriolimacidae). *Archiv für Molluskenkunde* 119:235–240.

———. 1992. The problem of the species in malacology after clear evidence of the limits of morphological systematics. In *Proceedings of the Ninth International Malacological Congress*, edited by E. Gittenberger and J. Goud, pp. 153–172.

———. 1997. How to distinguish *Oxychilus cellarius* (Müller, 1774) easily from *Oxychilus draparnaudi* (Beck, 1837) (Gastropoda, Pulmonata, Zonitidae). *Basteria* 61:43–56.

———. 2002. Redescription of two west European *Oxychilus* species: *O. alliarius* (Miller, 1822) and *O. helveticus* (Blum, 1881), and notes on the systematics of *Oxychilus* Fitzinger, 1833 (Gastropoda: Pulmonata: Zonitidae). *The Journal of Conchology* 37:455–476.

Godan, D. 1983. Pest slugs and snails: biology and control. Translated by Sheila Gruber. Berlin: Springer-Verlag; New York: Heidelberg.

Gómez, B.J. 2002. Structure and function of the reproductive system. In *The Biology of Terrestrial Gastropods*, edited by G.M. Barker., pp. 307–330. Wallingford, U.K.: CABI Publishing.

Goodfriend, G.A. 1986. Variation in land-snail shell form and size and its causes: a review. *Systematic Zoology* 35:204–223.

Groves, L.T. 1992. New range information for the bananaslug *Ariolimax columbianus* (Gould, 1851). *The Veliger* 35:157.

Guiller, A., M.A. Coutellec-Vreto, I. Madec, and J. Deunff. 2001. Evolutionary history of the land snail *Helix aspersa* in the Western Mediterranean: preliminary results inferred from mitochondrial DNA sequences. *Molecular Ecology* 10:81–87.

Gunn, A. 1992. The ecology of the introduced slug *Boettgerilla pallens* (Simroth) in North Wales. *The Journal of Molluscan Studies* 58:449–453.

Hand, C., and W.M. Ingram. 1950. Natural history observations on *Prophysaon andersoni* (J.G. Cooper), with special reference to amputation. *Bulletin, Southern California Academy of Sciences* 49:15–28.

Hanham, A.W. 1926. *Hemphillia malonei* Van[atta]. *The Nautilus* 39:43–44.

Hanna, G D. 1923. A new species of *Carychium* from Vancouver Island, British Columbia. *Proceedings of the California Academy of Sciences* (series 4) 12:51–53.

———. 1966. Introduced mollusks of western North America. *Occasional Papers of the California Academy of Science* 48.

Harris, S.A., and L. Hubricht. 1982. Distribution of the species of the genus *Oxyloma* (Mollusca, Succineidae) in southern Canada and the adjacent portions of the United States. *Canadian Journal of Zoology* 60:1607–1661.

Hausdorf, B. 1998. Phylogeny of the Limacoidea *sensu lato* (Gastropoda: Stylommatophora). *The Journal of Molluscan Studies* 64:35–66.

———. 2002. Introduced land snails and slugs in Colombia. *The Journal of Molluscan Studies* 68:127–131.

Heller, J. 1990. Longevity in Molluscs. *Malacologia* 31(2):259–295.

———. 1993. Hermaphroditism in molluscs. *Biological Journal of the Linnean Society* 48:19–42.

Heller, J., and A. Dolev. 1994. Biology and population dynamics of a crevice-dwelling landsnail, *Cristataria genezarethana* (Clausiliidae). *The Journal of Molluscan Studies* 60:33–46.

Heller, J., N. Sivan, and A.N. Hodgson. 1997. Reproductive biology and population dynamics of an ovoviviparous land snail, *Lauria cylindracea* (Pupillidae). *Journal of Zoology (London)* 243:263–280.

Henderson, J. 1927. Mollusk notes from the northwest. *The Nautilus* 40:75–78.

———. 1931. Molluscan provinces in the western United States. *The University of Colorado Studies* 18:177–186.

Hermida, J., P. Ondina, and A. Outeiro 1995. Ecological factors affecting the distribution of the gastropods *Aegopinella nitidula* (Draparnaud, 1805) and *Nesovitrea hammonis* (Ström, 1765) in northwest Spain. *The Journal of Conchology* 35:275–282.

Herzberg, F., and A. Herzberg. 1962. Observations on reproduction in *Helix aspersa*. *American Naturalist* 68:297–306.

Hoffmann, R.J. 1983. The mating system of the terrestrial slug *Deroceras laeve*. *Evolution* 37:423–425.

Hommay, G., F. Jacky, and M.F. Ritz. 1998. Feeding activity of *Limax valentianus* Férussac: nocturnal rhythm and alimentary competition. *The Journal of Molluscan Studies* 64:137–146.

Honek, A. 1995. Geographic distribution and shell colour and banding polymorphism in marginal populations of *Cepaea nemoralis* (Gastropoda, Helicidae). *Malacologia* 37:111–122.

Hubendick, B. 1950. The validity of *Vallonia excentrica* Sterki. *Proceedings of the Malacological Society of London* 28:75–78.

———. 1953. A second note on the validity of *Vallonia excentrica* Sterki. *Proceedings of the Malacological Society of London* 29:224–228.

Hubricht, L. 1985. The distributions of the native land mollusks of the eastern United States. *Fieldiana Zoology* (new series) 24.

Hudec, V. 1960. Critical evaluation of the species of the genus *Cochlicopa* Risso 1825 (Mollusca) found in Czechoslovakia. *Práce Brenskézákladny Ceskoslovenské Akademie Ved* 32:277–299.

Hyman, L.H. 1967. *The Invertebrates, Volume VI: Mollusca I.* New York: McGraw-Hill.

Ingram, W.M., and A.M. Peterson. 1947. Food of the giant western slug, *Ariolimax columbianus* (Gould). *The Nautilus* 61:49–51.

Jackson, R.W. 1923. *Epiphragmophora* a tree climber. *The Nautilus* 36:144.

Jones, J.S., B.H. Leith, and P. Rawlings. 1977. Polymorphism in *Cepaea*: a problem with too many solutions? *Annual Review of Ecology and Systematics* 8:109–143.

Jordaens, K, T. Backeljau, P. Ondina, H. Reise, and R. Verhagen. 1998. Allozyme homozygosity and phally polymorphism in the land snail *Zonitoides nitidus* (Gastropoda, Pulmonata). *Journal of Zoology (London)* 246:95–104.

Jordaens, K., S. Geenen, H. Reise, P. van Riel, R. Verhagen and T. Backeljau. 2000. Is there a geographical pattern in the breeding system of a complex of hermaphroditic slugs (Mollusca: Gastropoda: *Carinarion*)? *Heredity* 85:571–579.

Jordaens, K., P. van Riel, S. Geenen, R. Verhagen, and T. Backeljau. 2001. Food-induced body pigmentation questions the taxonomic value of colour in the self-fertilizing slug *Carinarion* spp. *The Journal of Molluscan Studies* 67:161–167.

Karlin, E.J. 1956. Notes on the ecology of *Zonitoides arboreus* (Say), *Opeas pumilum* (Pfeiffer) and *Lamellaxis gracilis* (Hulton) in greenhouses. *American Midland Naturalist* 55:121–125.

———. 1961. Ecological Relationships between vegetation and the distribution of land snails in Montana, Colorado and New Mexico. *American Midland Naturalist* 65:60–66.

Kerney, M. 1999. *Atlas of the land and freshwater molluscs of Britain and Ireland.* Colchester, U.K.: Harley Books.

Kerney, M.P., and R.A.D. Cameron. 1979. *A field guide to the land snails of Britain and North-West Europe.* London: Collins. (See also expanded editions: Kerney et al. 1983; Kerney and Cameron 1999.)

———. 1999. *Guide des Escargots et Limaces d'Europe.* Lausanne, France: Delcaux et Niestlé. Translated and adapted by A. Bertrand.

Kerney, M.P., R.A.D. Cameron, and J.H. Jungbluth. 1983. *Die Landschnecken Nord- und Mitteleuropas.* Hamburg: Paul Parey.

Kirchner, C., R. Kratzner and F.W. Welter-Schultes. 1997. Flying snails – how far can *Truncatellina* (Pulmonata: Vertiginidae) be blown over the sea? *The Journal of Molluscan Studies* 63:479–487.

Korte, A., and G.F.J. Armbruster. 2003. Apomorphic and plesiomorphic ITS-1 rDNA patterns in morphologically similar snails (Stylommatophora: Vallonia), with estimates of divergence time. *Journal of Zoology (London)* 260:275–283.

Kralka, R.A. 1986. Population characteristics of terrestrial gastropods in boreal forest habitats. *American Midland Naturalist* 115:156–164.

Kuiper, J.G.J. 1964. On *Vitrea contracta* (Westerlund). *The Journal of Conchology* 25:276–278.

Landolfa, M.A. 2002. On the adaptive function of the love dart of *Helix aspersa. The Veliger* 45:231–249.

La Rocque, A. 1953. Catalogue of the Recent Mollusca of Canada. *National Museum of Canada, Bulletin* 129.

———. 1962. Contributions to the history of Canadian malacology. *Sterkiana* 6:23–39.

Lebovitz, R.M. 1998. The inheritance of an embryonic lethal mutation in a self-reproducing terrestrial slug, *Deroceras laeve. Malacologia* 39:21–27.

Likharev, I., and E. Rammel'meier. 1962. Translated by Y. Lengy and Z. Krauthamer. *Terrestrial mollusks of the fauna of the U.S.S.R.* (Keys to the Fauna of the U.S.S.R.) Jerusalem: Israel Program for Scientific Translation.

Lissmann, H.W. 1945. The mechanism of motion in gastropod molluscs. II. Kinetics. *Journal of Experimental Biology* 22:37–50.

Lloyd, D.C. 1970a. The function of the odour of the garlic snail *Oxychilus alliarius* (Pulmonata: Zonitidae). *Malacologia* 10:441–449.

———. 1970b. The composition of the odour of the garlic snail *Oxychilus alliarius* (Pulmonata: Zonitidae). *Malacologia* 10:451–455.

———. 1970c. The use of skin characters as an aid to the identification of the British species of *Oxychilus* (Fitzinger) (Mollusca, Pulmonata, Zonitidae). *Journal of Natural History* 4:531–534.

Lohmander, H. 1938. Landmollusken aus Island, gesammelt von Dr Carl H. Lindroth, 1929. *Meddelanden från Göteborgs Musei Zoologiska Avdelning*, 76. *Göteborgs Kungl. Vetenskaps-och Vitterhets-Samhälles handlingar*, 5B, 6:52, pl. 4.

Lord, J.K. 1866. *The Naturalist in Vancouver Island and British Columbia*. 2 vols. London: Richard Bentley.

McCracken, G.F., & R.K. Selander. 1980. Self-fertilization and monogenic strains in natural populations of terrestrial slugs. *Proceedings of the National Academy of Sciences of the U.S.A.* 77:684–688.

McGraw, R., N. Duncan, and E. Cazares. 2002. Fungi and other items consumed by the blue-gray taildropper slug (*Prophysaon coeruleum*) and the papillose taildropper slug (*Prophysaon dubium*). *The Veliger* 45:261–264.

Mead, A.R. 1943. Revision of the giant West Coast land slugs of the genus *Ariolimax* Moerch (Pulmonata: Arionidae). *The American Midland Naturalist* 30:675–713.

Metcalf, A.L., and R.A. Smartt. 1997. Land snails of New Mexico: a systematic review. In *Land snails of New Mexico*, edited by A.L. Metcalf and R.A. Smartt. *New Mexico Museum of Natural History and Science, Bulletin* 10:1–69.

Moens, R. 1990. The predation of *Zonitoides nitidus* on eggs of gastropods: relationship between soil moisture and vegetation cover. *Journal of Medical and Applied Malacology* 1:107–112.

Mordan, P.B. 1977. Factors affecting the distribution and abundance of *Aegopinella* and *Nesovitrea* (Pulmonata: Zonitidae) at Monks Wood National Nature Reserve, Huntingdonshire. *Biological Journal of the Linnean Society* 9:59–72.

———. 1978. The life cycle of *Aegopinella nitidula* (Draparnaud) (Pulmonata: Zonitidae) at Monks Wood. *The Journal of Conchology* 29:247–252.

Morton, J.E. 1955. Notes on the ecology and annual cycle of *Carychium tridentatum* at Box Hill. *Proceedings of the Malacological Society of London* 31:30–46.

———. 1979. *Molluscs*, 5th edition. London: Hutchinson.

Myers, L.D. 1972. Primary and secondary influencing agents on gastropod populations of three habitats in Washington State. *Sterkiana* 47:39–45.

Nekola, J.C. 2001. Distribution and ecology of *Vertigo cristata* Sterki, 1919 in the Western Great Lakes Region. *American Malacological Bulletin* 16:47–52.

Nekola, J.C., and M. Barthel. 2002. Morphometric analysis of *Carychium exile* and *Carychium exiguum* in the Great Lakes region of North America. *The Journal of Conchology* 37:515–531.

Nekola, J.C., and P.A. Massart. 2001. Distribution and ecology of *Vertigo nylanderi* Sterki, 1909 in the western Great Lakes region. *American Malacological Bulletin* 16:53–60.

Nicklas, N.L., and R.J. Hoffmann. 1981. Apomictic parthenogenesis in a hermaphroditic terrestrial slug, *Deroceras laeve* (Müller). *Biological Bulletin* 160:123–35.

Oughton, J. 1948. A zoogeographical study of the land snails of Ontario. *University of Toronto Studies, Biological Series* 57.

Outeiro, A., S. Mato, I. Riballo, and T. Rodriguez. 1990. On *Cochlicopa lubrica* (Müller, 1774) and *Cochlicopa lubricella* (Porro, 1837) (Gastropoda: Pulmonata: Cochlicopidae) in the Sierra de O Courel (Lugo, NW Spain). *The Veliger* 33:408–415.

Ovaska, K., L. Chichester, H. Reise, W.P. Leonard, and J. Baugh. 2002. Anatomy of the dromedary jumping-slug, *Hemphillia dromedarius* Branson,

1972 (Gastropoda: Stylommatophora: Arionidae), with new distributional records. *The Nautilus* 116:89–94.

Ovaska, K., W.P. Leonard, L. Chichester, T.E. Burke, L. Sopuck, and J. Baugh. In press. *Prophysaon coeruleum* Cockerell, 1890, Blue-Gray Taildropper (Gastropoda: Arionidae): new distributional records and reproductive anatomy. *Western North American Naturalist.*

Pakarinen, E. 1994a. Autotomy in arionid and limacid slugs. *The Journal of Molluscan Studies* 60:19–23.

———. 1994b. The importance of mucus as a defence against carabid beetles by the slugs *Arion fasciatus* and *Deroceras reticulatum. The Journal of Molluscan Studies* 60:149–155.

Peake, J. 1978. Distribution and ecology of the Stylommatophora. In: *Pulmonates. Volume 2A, Systematics, Evolution and Ecology,* pp. 429–526. London: Academic Press.

Pfeiffer, L. 1850. Descriptions of twenty-four new species of Helicea, from the collection of H. Cuming, Esq. *Proceedings of the Zoological Society, London for 1849,* 17:126–131.

Pillard, D.A. 1985. Mucus trail following by the slug *Deroceras laeve* (Müller). *Journal of the Tennessee Academy of Science* 60:13–15.

Pilsbry, H.A. 1898. A classified list of American land shells, with localities. *The Nautilus* 11:117–120.

———. 1939–1948. Land Mollusca of North America (north of Mexico). *The Academy of Natural Sciences of Philadelphia, Monograph 3.* 1939: 1(1):i–xvii, 1–573, i–ix. 1940: 1(2):575–994, i–ix. 1946: 2(1):i–iv, 1–520, i–ix. 1948: 2(2):i–xlvii, 521–1113.

———. 1953. *Magnipelta,* a new genus of Arionidae from Idaho. *The Nautilus* 67:37–38.

Pilsbry, H.A., and R.B. Brunson. 1954. The Idaho-Montana slug *Magnipelta* (Arionidae*). Notulae Naturae of the Academy of Natural Sciences of Philadelphia* 262.

Pilsbry, H.A., and C.M. Cooke. 1922. Land shells of Vancouver Island. *The Nautilus* 36:37–38

Pinceel, L., K. Jordaens, P. van Riel, G. Bernon, and T. Backeljau. 2002. Molecular characterization of an alien slug in North America. Abstracts, 9th Benelux Congress of Zoology, "Adaptation and Constraint", University of Antwerp, Belgium, 8–9 November 2002: 167.

Platt, T.R. 1980. Observations on the terrestrial gastropods in the vicinity of Jasper, Alberta. *The Nautilus* 94:18–21.

Plisetskaya, E.M., and I. Deyrup-Olsen. 1987. An insulin-like substance in the blood of the slug *Prophysaon foliolatum* (Arionidae) in the course of tail regeneration. *Comparative Biochemistry and Physiology: A Comparative Physiology* 87:781–783.

Pokryszko, B.M. 1987a. European *Columella* reconsidered (Gastropoda: Pulmonata: Vertiginidae). *Malakologische Abhandlungen Staatliches Museum für Tierkunde Dresden* 12:1–12.

———. 1987b. On the aphally in the Vertiginidae (Gastropoda: Pulmonata: Orthurethra). *The Journal of Conchology* 32:356–375.

———. 1990. The Vertiginidae of Poland (Gastropoda: Pulmonata: Pupilloidea) – a systematic monograph. *Polska Akademia Nauk Instytut Zoologii, Annales Zoologici* 43(8):133–257.

Pokryszko, B.M. 1997. Land snail apertural barriers – adaptation or hindrance? (Gastropoda: Pulmonata). *Malakologische Abhandlungen Staatliches Museum für Tierkunde Dresden* 18:239–248.

Ponder, W.F., and D.R. Lindberg. 1996. Gastropod phylogeny – changes for the 90s. In *Origin and evolutionary radiation of the Mollusca*, edited by J. Taylor, pp. 135–154. Oxford: Oxford University Press.

Porter, C.A. 1965. Comparison of the genitalia of two sympatric species of *Haplotrema*. *The Nautilus* 79:19–23.

Ports, M.A. 1996. Habitat affinities and distributions of land gastropods from the Ruby Mountains and East Humboldt Range of northeastern Nevada. *The Veliger* 39:335–341.

Quick, H.E. 1947. *Arion ater* (L.) and *A. rufus* (L.) in Britain and their specific differences. *The Journal of Conchology* 22:249–261.

————. 1954. *Cochlicopa* in the British Isles. *Proceedings of the Malacological Society, London* 30:204–213.

————. 1960. British slugs (Pulmonata: Testacellidae, Arionidae, Limacidae). *Bulletin of the British Museum (Natural History), Zoology* 6:103–226.

Randolph, P.B. 1899. Collecting shells in the Klondike country. *The Nautilus* 12:109–112.

Rees, B.B. 1988. Electrophoretic and morphological characteristics of two species of *Oreohelix*, the mountain snail. *Malacological Review* 21:129–132.

Rees, W.J. 1965. The aerial dispersal of Mollusca. *Proceedings of the Malacological Society of London* 36:269–289.

Reise, H., J.M.C. Hutchinson, R.G. Forsyth, and T.J. Forsyth. 2000. The ecology and rapid spread of the terrestrial slug *Boettgerilla pallens* in Europe with reference to its recent discovery in North America. *The Veliger* 43:313–318.

Reise, H., B. Zimdars, K. Jordaens, and T. Backeljau. 2001. First evidence of possible outcrossing in the terrestrial slug *Arion intermedius* (Gastropoda: Pulmonata). *Hereditas* 134:267–270.

Richardson, A.M.M. 1975. Food, feeding rate and assimilation in the landsnail, *Cepaea nemoralis*. *Oecologia* 19:59–70.

Richter, K.O. 1979. Aspects of nutrient cycling by *Ariolimax columbianus* in Pacific Northwest coniferous forests. *Pedobiologia* 19:60–74.

————. 1980. Movement, reproduction, defence, and nutrition as functions of the caudal mucus in *Ariolimax columbianus*. *The Veliger* 23:43–47.

Riedel, A. 1957. Revision der Zonitiden Polens (Gastropoda). *Annales Zoologici, Warszawa* 16:361–461.

————. 1983. Über die *Aegopinella*-Arten (Gastropoda, Zonitidae) aus Jugoslawien, Italien und Frankreich. *Annales Zoologici, Warszawa* 37:235–258.

Rigby, J.E. 1963. Alimentary and reproductive systems of *Oxychilus cellarius* (Müller) (Stylommatophora). *Proceedings of the Zoological Society of London* 141:311–359.

Robinson, D.G. 1999. Alien invasions: the effects of the global economy on non-marine gastropod introductions into the United States. *Malacologia* 41:413–438.

Rollo, C.D., and W.G. Wellington. 1975. Terrestrial slugs in the vicinity of Vancouver, British Columbia. *The Nautilus* 89:107–115.

————. 1981. Environmental orientation by terrestrial Mollusca with particular reference to homing behavior. *Canadian Journal of Zoology* 59:225–239.

Roscoe, E.J. 1962. Aggregations of the terrestrial pulmonate *Cionella lubrica. The Nautilus* 75:111–115.

Roth, B. 1977. *Vitrea contracta* (Westerlund) (Mollusca: Pulmonata) in the San Francisco Bay area. *The Veliger* 19:429–430.

———. 1981. Shell color and banding variation in two coastal colonies of *Monadenia fidelis* (Gray) (Gastropoda: Pulmonata). *The Wasmann Journal of Biology* 38:39–51.

———. 1982a. European land mollusks in the San Francisco Bay area, California: *Carychium minimum* Müller and the *Arion hortensis* complex. *The Veliger* 24:342–344.

———. 1982b. *Discus rotundatus* (Müller) (Gastropoda: Pulmonata) in California. *Malacological Review* 15:139–140.

———. 1985. A new species of *Punctum* (Gastropoda: Pulmonata: Punctidae) from the Klamath Mountains, California, and first Californian records of *Planogyra clappi* (Valloniidae). *Malacological Review* 18:51–56.

———. 1986. Notes on three European land mollusks introduced to California. *Bulletin of the Southern California Academy of Sciences* 85:22–28.

———. 1987a. Identities of two Californian land mollusks described by Wesley Newcomb. *Malacological Review* 20:129–132.

———. 1987b. "*Punctum pusillum*" – a correction. *The Veliger* 30:95–96.

———. 1990. New haplotrematid land snails, *Ancotrema* and *Haplotrema* (Gastropoda: Pulmonata), from the Sierra Nevada and North Coast Ranges, California. *The Wasmann Journal of Biology* 47:68–76.

———. 1991. A phylogenetic analysis and revised classification of the North American Haplotrematidae (Gastropoda: Pulmonata). *American Malacological Bulletin* 8:155–163.

———. 2004. Observations on the taxonomy and range of *Hesperarion* Simroth, 1891 and the evidence for genital polymorphism in *Ariolimax* Mörch, 1860 (Gastropoda: Pulmonata: Arionidae: Ariolimacinae). *The Veliger* 47:38–46.

Roth, B., and D.R. Lindberg. 1981. Terrestrial mollusks of Attu, Aleutian Islands, Alaska. *Arctic* 34:43–47.

Roth, B., and W.B. Miller. 1993. Polygyrid land snails, *Vespericola* (Gastropoda: Pulmonata), 1. Species and populations formerly referred to *Vespericola columbianus* (Lea) in California. *The Veliger* 36:134–144.

Roth, B., and T.A. Pearce. 1984. *Vitrea contracta* (Westerlund) and other introduced land mollusks in Lynnwood, Washington. *The Veliger* 27:90–92.

Roth, B., and P.S. Sadeghian. 2003. Checklist of the land snails and slugs of California. Santa Barbara Museum of Natural History, *Contributions in Science* 3.

Runham, N.W., and A.A. Laryea. 1968. Studies on the maturation of the reproductive system of *Agriolimax reticulatus* (Pulmonata: Limacidae). *Malacologia* 7:93–108.

Russell, L.S. 1951. Land snails of the Cypress Hills and their significance. *The Canadian Field-Naturalist* 65:174–175.

Schileyko, A.A. 2002. Treatise on recent terrestrial pulmonate molluscs. Part 8. Punctidae, Helicodiscidae, Discidae, Cystopeltidae, Euconulidae, Trochomorphidae. *Ruthenica* (Supplement 2):1035–1166.

Seddon, M.B., and D.T. Holyoak. 1993. Land Gastropods of NW. Africa: new distributional data and nomenclature. *The Journal of Conchology* 34:321–331.

Shen, J. 1995. Cannibalism in the terrestrial slug *Deroceras laeve. The Nautilus* 109:41–42.

Singh, S.M. 1981. Polymorphism in colonies of the land snail *Cepaea nemoralis* at London, Ontario: changes over three decades. *The Canadian Field-Naturalist* 95:192–197.

Smith, A.G. 1943. Mollusks of the Clearwater Mountains, Idaho. *Proceedings of the California Academy of Sciences* (series 4) 23:537–554.

Smith, A.G., W.B. Miller, C.C. Christensen, and B. Roth. 1990. Land Mollusca of Baja California, Mexico. *Proceedings of the California Academy of Sciences* (series 4) 47:95–158.

Smith, B.J., and R.C. Kershaw. 1979. *Field guide to the non-marine molluscs of south eastern Australia*. Canberra: Australian National University Press.

Solem, A. 1972. Microarmature and barriers in the apertures of land snails. *The Veliger* 15:81–87.

———. 1974. *The Shell Makers: Introducing Mollusks*. New York: John Wiley & Sons.

———. 1975. Notes on Salmon River valley oreohelicid land snails, with description of *Oreohelix waltoni*. *The Veliger* 18:16–30.

———. 1977a. *Radiodiscus hubrichti* Branson, 1975, a synonym of *Striatura pugetensis* (Dall, 1895) (Pulmonata: Zonitidae). *The Nautilus* 91:146–148.

———. 1977b. Shell microsculpture in *Striatura, Punctum, Radiodiscus*, and *Planogyra* (Pulmonata). *The Nautilus* 91:149–155.

———. 1984. A world model of land snail diversity and abundance. In *World-wide snails: biogeographical studies on non-marine Mollusca*, edited by A. Solem and A.C. van Bruggen, pp. 6–22. Leiden: E.J. Brill.

South, A. 1965. Biology and ecology of *Agriolimax reticulatus* (Müller) and other slugs: spatial distribution. *Journal of Animal Ecology* 34:403–417.

———. 1992. *Terrestrial slugs: biology, ecology and control*. London: Chapman & Hall.

Sparks, M.A. 1953. Fossil and Recent English species of *Vallonia*. *Proceedings of the Malacological Society of London* 30:110–121.

Stanisic, J. 1981. *Helix parramattensis* Cox, 1864: A synonym of *Euconulus (Euconulus) fulvus* (Muller, 1774) (Pulmonata: Euconulidae). *Journal of the Malacological Society of Australia* 5:81–83.

Starobogatov, Y.I. 1996. Eurasiatic species of the genus *Cochlicopa* (Gastropoda, Pulmonata, Cochlicopidae). *Ruthenica* 5:105–129.

Stasek, C.R. 1967. Autotomy in the Mollusca. *Occasional Papers of the California Academy of Sciences* 61.

Stephenson, J.W. 1979. The functioning of the sense organs associated with feeding behaviours in *Deroceras reticulatum* (Müll.). *The Journal of Molluscan Studies* 45:167–171.

Taylor, G.W. 1889. The land shells of Vancouver Island. *The Ottawa Naturalist* 3:84–94.

———. 1891a. Land shells of Vancouver Island. *The Nautilus* 5:91.

———. 1891b. *Limax agrestis* Linn. on the Pacific Coast. *The Nautilus* 5:92–93.

———. 1900. *Hemphillia glandulosa*. A slug new to the Canadian list. *The Ottawa Naturalist* 14:150–151.

Taylor, J.W. 1906–1914. *A Monograph of the Land and Freshwater Mollusca of the British Isles*, Vol. 3 (Parts 12–21). Leeds, U.K.: Taylor Brothers.

Tompa, A.S. 1979. Oviparity, egg retention and ovoviviparity in pulmonates. *The Journal of Molluscan Studies* 45:155–160.

———. 1984. Land snails (Stylommatophora). In *The Mollusca. Volume 7,*

Reproduction, edited by A.S. Tompa, N.H. Verdonk and J.A.M. van den Biggelaar, pp 47–140. New York: Academic Press.

Trueman, E.R. 1983. Locomotion in molluscs. In *The Mollusca*, Vol. 4: *Physiology*, Part 1, edited by S.M. Saleuddin and K.M. Wilbur. New York: Academic Press.

Turgeon, D.D., J.F. Quinn, A.E. Bogan, E.V. Coan, F.G. Hochberg, W.G. Lyons, P.M. Mikkelsen, R.J. Neves, C.F.E. Roper, G. Rosenberg, B. Roth, A. Scheltema, F.G. Thompson, M. Vecchione and J.D. Williams. 1998. Common and scientific names of aquatic invertebrates from the United States and Canada: Mollusks. 2nd ed. *American Fisheries Society Special Publication* 26.

Tuthill, S.J., and R.L. Johnson. 1969. Nonmarine mollusks of the Katalla Region, Alaska. *The Nautilus* 83:44–52.

Vagvolgyi, J. 1968. Systematics and evolution of the genus *Triodopsis* (Mollusca: Pulmonata: Polygyridae). *Bulletin of the Museum of Comparative Zoology* 136:145–254.

Valovirta, I., and R.A. Väisänen. 1986. Multivariate morphological discrimination between *Vitrea contracta* (Westerlund) and *V. crystallina* (Müller) (Gastropoda, Zonitidae). *The Journal of Molluscan Studies* 52:62–67.

Vanatta, E.G. 1906. British Columbia shells. *The Nautilus* 20:95.

Van Goethem, J. 1972. Contribution a l'Etude de *Boettgerilla vermiformis* Wiktor, 1959 (Mollusca, Pulmonata). *Bulletin de l'Institute Royale des Sciences Naturelles de Belgique* 48:1–16.

Vaught, K.C. 1989. *A Classification of the Living Mollusca*. Melbourne, Florida: American Malacologists.

von Proschwitz, T. 1985. Die Landschneckenfauna des Ammarnäs-Gebietes (Lappland, Nordschweden), mit einigen Bemerkungen zur Gattung *Euconulus* Reinh. (Gastropoda). *Malakologische Abhandlungen Staatliches Museum für Tierkunde Dresden* 10:95–108.

———. 1994. *Oxychilus cellarius* (Müller) and *Oxychilus draparanaudi* (Beck) as predators on egg-clutches of *Arion lusitanicus* Mabille. *The Journal of Conchology* 35:183–184.

Wade, C.M., and P.B. Mordan. 2000. Evolution within the gastropod Mollusca; using the ribosomal RNA gene-cluster as an indicator of phylogenetic relationships. *The Journal of Molluscan Studies* 66:565–570.

Wade, C.M., P.B. Mordan, and B. Clarke. 2001. A phylogeny of the land snails (Gastropoda: Pulmonata). *Proceedings of the Royal Society of London B*, 268:413–422.

Waldén, H.W. 1962. On the variation, nomenclature, distribution, and taxonomic position of *Limax (Lehmannia) valentiana* Férussac (Gastropoda, Pulmonata). *Arkiv för Zoologi*, Serie 2, 15:71–96, pl. 1.

Walton, M.L 1970. Longevity in *Ashmunella, Monadenia* and *Sonorella*. *The Nautilus* 83:109–112.

Watabe, N. 1988. Shell structure. In *The Mollusca*, Vol. 11: *Form and Function*, edited by E.R. Trueman and M.R. Clarke, pp. 69–104. New York: Academic Press.

Watson, H. 1920. The affinities of *Pyramidula, Patulastra, Acanthinula,* and *Vallonia*. *Proceedings of the Malacological Society* 14:6–30, 2 pls.

———. 1923. Masculine deficiencies in British Vertiginidae. *Proceedings of the Malacological Society of London* 15:270–280.

———. 1934. Genital dimorphism in *Zonitoides*. *Journal of Conchology* 20:34–42.

Watson, H., and B. Verdcourt. 1953. The two British species of *Carychium*. *The Journal of Conchology* 23:306–324.

Webb, G.R. 1952. Pulmonata. Xanthonycidae: Comparative sexological studies of the North American land-snail *Monadenia fidelis* (Gray). *Gastropodia* 1:7–8.

———. 1961. The sexology of three species of limacid slugs. *Gastropodia* 1:53–55.

Webb, G.R., and R.H. Russell. 1977. Anatomical notes on a *Magnipelta*: Camaenidae? *Gastropodia* 1:107–108.

Westfall, J.A. 1959. Oviposition, hatching and early development in *Ariolimax columbianus* (Gould). *The Veliger* 2:10–11.

Wiktor, A. 2000. Agriolimacidae (Gastropoda: Pulmonata) – a systematic monograph. *Annales Zoologici* 49:347–590.

Williamson, P. 1979. Age determination of juvenile and adult *Cepaea*. *The Journal of Molluscan Studies* 45:52–60.

Williamson, P., and R.A.D. Cameron. 1976. Natural diet of the landsnail *Cepaea nemoralis*. *Oikos* 27:493–500.

Wolda, H. 1970. Variation in growth rate in the land snail *Cepaea nemoralis*. *Researches in Population Ecology* 12:185–204.

Wolda, H.P., A. Zweep, and K.A. Schuitema. 1971. The role of food in dynamics of populations of the land snail *Cepaea nemoralis*. *Oecologia* 7:361–381.

Yoon, S., and W. Kim. 2000. Phylogeny of some gastropod mollusks derived from the 18S rDNA sequences with emphasis on the Euthyneura. *The Nautilus* 114:84–92.

GLOSSARY

Aestivate   To be dormant during dry conditions, especially during the summer.

Apertural lip   The edge of the aperture (figures 11, 14).

Aperture   The opening in a spirally coiled shell through which the animal extends and retracts; also called "mouth".

Apex (adj: apical)   The tip of the spire, often pointed; the part of the shell that is formed first (figure 11).

Aphally (adj: aphallic)   The condition of the genital tract when male sexual organs are absent. See also *euphally, hemiphally*.

Atrium   The portion of the reproductive system connected to the outside by way of the genital pore, into which the oviduct, penis (or epiphallus) and bursa copulatrix duct connect (figure 5).

Autotomy   The voluntary casting off of a body part; in slugs, casting off part of the tail when under attack.

Axial   In the same direction as the axis of a coiled shell.

Axis   The imaginary line around which the whorls of a coiled shell are formed.

Bursa copulatrix   A sac-like part of the genital system that receives excess spermatozoa and other reproductive products for reabsorption (figure 5); sometimes called "seminal receptacle" or "spermatheca".

Bursa copulatrix duct   A tube connecting the *bursa copulatrix* to the rest of the genital system (figure 5); also called "seminal" or "spermathecal" duct.

Caecum   A blind duct or pocket on the rectum (a portion of the intestinal system).

Calciphile   An organism associated with limestone and calciferous soils, and requiring large amounts of free calcium ions.

Callus   A thick, often opaque shell deposit; most calluses on a snail shell occur inside the apertural lip (see *palatal callus*).

Caudal mucus plug   A dried piece of mucus in the caudal mucus pore.

Caudal mucus pore   A mucus-secreting gland at the end of the tail in some slugs.

Columella (adj: columellar)   The central pillar formed by the inner walls of the whorls, through which the axis of the shell passes, either hollow (with an umbilicus) or solid.

Columellar baffle   A more-or-less vertical, denticle-like projection hidden well behind the columella in some species of Pupillidae (figure 15).

Crest   A raised, axial ridge on the last whorl behind, and parallel to, the apertural lip, typically set off from the lip by a constriction; it is present in some Pupillidae, Vertiginidae and Polygyridae (figure 15).

Dentate   Having denticles.

Denticle   A small tooth-like projection on the *apertural lip* or behind the lip and farther back within the shell; also called "tooth", "lamella" or "lamina". See figure 15 for the names of individual denticles.

Denticulate   Finely toothed.

Depressed-heliciform   Heliciform, with a lower spire (figure 10).

Distal   Located away from the centre of the body or point of attachment. Here, the distal genitalia is the part of the reproductive system closest to the genital pore.

Epiphallus   A dilated distal portion of the vas deferens (figure 5) in which the spermatophore is secreted.

Epiphragm   A mucus sheet that is secreted to seal the shell aperture.

Euphally (adj: euphallic)   The "normal" condition of the genital tract where both male and female sexual organs are fully developed. See also *aphally, hemiphally*.

False keel   A row of tubercles that form the appearance of a keel on the tail of some *Arion* slugs.

Foot   The muscular organ of gastropod locomotion.

Foot fringe   The edge of the foot, when set off from the rest of the foot by a conspicuous constriction; sometimes called the "skirt".

Genital pore   The opening or openings on the anterior right side of the body leading into the reproductive system; sometimes called the "genital aperture".

Granular   Covered with *granules*.

Granule   A small raised surface feature somewhat like a grain of sand.

Height (of a shell)   The maximum measurement taken along the line of axis from the apex to the base or basal lip (figure 18); sometimes called "length".

Heliciform   The typical shape of a land snail's shell, like that of *Cornu* species (figure 10).

Hemiphally (adj: hemiphallic)   The condition of the genital tract when the female sexual organs are fully developed and the male organs only partially developed.

Holarctic   The biogeographic region consisting of both the *Palearctic* and *Nearctic*; an organism native to the Holarctic.

Incremental striae   Sculptural elements on the surface of a shell (see figure 17), usually fine, that are created by the gradual addition of shell material as a result of growth; sometimes called "axial", "radial" or "growth" striae/lines.

Keel   The longitudinal ridge on the tail of some slugs (figure 4).

Lamella (pl: lamellae)   A thin, plate-like element.

Lamellar   Thin and flat, like a lamella.

Lip   See *apertural lip*.

Mantle   A fold of the body wall that lines and secretes the shell in shell-bearing molluscs; it is exposed in slugs (figure 4).

Nearctic   The biogeographic region including all of North America and northern, non-tropical Mexico; an organism native to the region.

Of authors   Indicates that a name is used by subsequent authors in a different sense than intended by the original author of a species (or genus). It implies a misuse or misidentification by later authors.

Outcrossing   The production of offspring by the fusion of gametes from different individuals.

Ovate   Egg-shaped.

Oviduct   The female duct through which eggs pass from the spermoviduct to the atrium (figure 5).

Oviparous   Producing young that develop inside an egg after it is laid.

Ovoviviparous   Producing young that develop in an egg retained inside the parent's body.

Palearctic   The biogeographic region including Europe, northern Asia and North Africa; an organism native to the Palearctic region.

Palatal callus   A thick, opaque ridge inside the aperture, back from the lip (figure 15), on which palatal denticles sit.

Papillae   A small dermal projection similar in form to a nipple.

Parietal   Of the inner wall of a coiled shell between columella and the suture, formed by the preceding whorl uniting with the apertural lip.

Penial flagellum (pl: flagellae)   An outgrowth from the proximal part of the penis; also called "appendix" or "penial caecum".

Periostracum   A thin chitinous covering on the exterior of many shells, often with hairs or wrinkles.

Periphery   The part of the whorl farthest from the *axis* of a spiral shell (figure 16).

Pneumostome   A hole on the right side of the *mantle* through which the animal breathes (figures 3, 4); also called "breathing pore".

Protoconch   The initial *whorl* or whorls, usually characterized by sculptural differences from the rest of the shell.

Proximal   Located near the centre of the body or point of attachment.

Recurved   Curved or bent back, as at the edge of the apertural lip (figure 14); sometimes called "reflected".

Radula   A feeding organ of most molluscs, generally consisting of rows of minute teeth on a tongue-like membrane.

Rib   A long, narrow ridge-like element of surface ornamentation (figure 17); sometimes called "costa".

Self-fertilization   Fertilization of an egg by sperm from the same individual.

Shoulder   That part of the whorl directly below the suture and above the periphery, either angular or rounded.

Sinulus   An indentation of the apertural lip in some species of Vertiginidae and Pupillidae (figure 15); sometimes called "auricle".

Spiral   In the direction of the coiling of the shell.

Spire   All *whorls* of a coiled shell except for the last.

Subcylindrical   Approximately cylindrical (figure 10).

Subovate   Approximately egg-shaped (figure 10).

Succineid   A member of the family Succineidae.

Suture   The continuous seam between two adjacent whorls of a coiled shell (figure 11).

Synanthrope (adj: synanthropic)   An organism associated with humans or human dwellings.

Sympatric   Living together in the same area.

Tentacles   Sensory organs on the head of a snail or slug, slender and either contractile or retractile (figures 2, 3, 4).

Thread   A narrow, raised, usually *spiral* sculptural element (figure 17).

Tubercle   A swelling, hump or knob.

Umbilicus   In some coiled shells, the hole, indentation or depression formed when the inner surfaces of the whorls do not join (figure 12).

Vas deferens   The narrow duct connecting the spermoviduct to either the penis or epiphallus (figure 5).

Whorl   A full coil of the tube of a snail's shell (figure 19).

Width (of a shell)   The maximum measurement taken perpendicular to the axis; sometimes called "breadth" or "diameter".

Scientific Names of Plants and Other Animals Mentioned in This Book

Balsam Poplar	*Populus balsamifera balsamifera*
Bigleaf Maple	*Acer macrophyllum*
Black Cottonwood	*Populus balsamifera trichocarpa*
Devil's Club	*Oplopanax horridus*
Douglas-fir	*Pseudotsuga menziesii*
English Ivy	*Hedera helix*
Garry Oak	*Quercus garryana*
Leach's Storm Petrel	*Oceanodroma leucorhoa*
Skunk Cabbage	*Lysichiton americanum*
Spreading Phlox	*Phlox diffusa*
Stinging Nettle	*Urtica dioica*
Sword Fern	*Polystichum munitum*
Trembling Aspen	*Populus tremuloides*

ACKNOWLEDGEMENTS

I thank the many people who have generously helped with this book. Most of the photographs of the living animals were taken by Dr Kristiina Ovaska (Victoria) and most of the drawings of introduced slugs and slug anatomies are the fine work of Dr Andrzej Wiktor (Museum of Natural History, Wroclaw, Poland). All photographs and illustrations not my own are acknowledged on the following page.

For providing literature or helping on specific questions, I thank Dr Georg Armbruster (University of Basel, Switzerland), Dr R.A.D. Cameron (University of Sheffield, England), Dr Brian Coles (Jefferson, Arkansas), Dr Derek Davis, (Nova Scotia Museum of Natural History, Halifax), Neil Fahy (California Academy of Sciences, San Francisco), Dr Terrence Frest (Seattle), Wayne Grimm (Kemptville, Ontario), Dr Stuart Harris (University of Calgary), Dr Jeff Nekola (University of Wisconsin – Green Bay) and Dr Barry Roth (San Francisco). On numerous occasions, Dr Heike Reise (Staatliches Museum für Naturkunde Görlitz, Germany) and Dr John Hutchinson (Berlin) offered advice, literature and specimens. Larry Williams (Burnaby, B.C.) made useful suggestions for the keys. Elizabeth Kools (California Academy of Sciences, San Francisco), Dr Jean-Marc Gagnon (Canadian Museum of Nature, Ottawa), Dr Jochen Gerber (Field Museum of Natural History, Chicago) and Dr Guenter Schuster (Branley A. Branson Museum of Zoology, Eastern Kentucky University), and their respective institutions, helped by answering my queries and making specimens available for loan. And I thank Drs Roth, Reise, Hutchison and Harris again for their careful and constructive reviews of the manuscript.

My wife, Tammy Forsyth, was in many ways instrumental in the completion of the manuscript. Her eagerness to explore new territory in search of terrestrial molluscs laid the foundation for this book.

I thank Philip Lambert (Curator of Invertebrates), Kelly Sendall (Collections Manager of Invertebrates, Fish and Herpetology), Gerry Truscott (Publisher), and other Royal BC Museum staff who supported me during the writing of *Land Snails of British Columbia*.

Land Snails of British Columbia

Edited, designed and typeset by Gerry Truscott, RBCM.
The body type is Palatino 10/12; the headings are also in Palatino.
The figure captions and running footers are in Optima.

Cover design by Chris Tyrrell, RBCM.

Printed in Canada by Kromar Printing, Winnipeg.

SPECIES INDEX

Items are listed by genus and species names; some common names are included. Pages in bold indicate the main entry.

Aegopinella nitidula 20, 26, 98, 99, **101-2**, 104
agna, Bifidaria 53
Agriolimax agrestis 121; *campestris* 118;
    caruanae 120; *hemphilli* 118;
    montanus 118
alaskana, Vitrina 109, 110
alaskensis, Conulus fulvus 94;
    Euconulus fulvus 95
albescens, Pupilla hebes 51
albula, Vallonia 49, 50
alderi, Euconulus 96; *Helix fulva* 95
alleni, Punctum conspectum 75
alliaria, Helix 104
alliarius, Oxychilus 11, 27, **104-5**, 106, C-24
Allogona lombardii **156**;
    ptychophora 18, 28, 79, **148-49**, 149;
    townsendiana 18, 28, 148, **149-50**, C-29
alticola, Columella 55; *Pupilla* 54
Ancotrema hybridum 25, **70-71**, 71, 72, C-33;
    portella 25, 71, **71-72**
andersoni, Arion 142;
    Prophysaon 9, 19, 28, **142-43**, 145, C-26
andrusiana, Vertigo 25, **59-60**, 61
angelicae, Vitrina 109
Anguispira kochi 18, 25, **79-80**, C-1
anthonyi, Pyramidula cronhitei 81
antiquorum, Pupa ovata 67
apiarium, Oreohelix cooperi 85
arborea, Helix 98
arboreus, Zonitoides 18, 26, **98-99**, 99, 101, 105
arctica, Hyalina 91; *Vertigo* 66
arcticum, Pristiloma 26, 90, **91-92**
Ariolimax columbianus 10, 11, 12, 18, 27, **123-24**,
    C-16-18; *steindachneri* 123
Arion andersoni 142; *ater* 127, 128, **147**;
    circumscriptus 27, **129-30**, 130, 131, 147, C-7;
    distinctus 27, **132-33**, 148, C-6;
    fasciatus 129, 130, 131, **147**; *foliolatus* 145;
    hortensis 132, 133, **148**; *intermedius* 27,
    128, **133-34**, C-11; *owenii* 133
rufus 27, **127-28**, 133, 147, C-8, C-9;
    silvaticus 27, 130, **130-31**;
    subfuscus 4, 7, 28, **135-36**, C-10
arthuri, Vertigo 25, **60-61**, 64, 65
ashmuni, Agriolimax hemphilli 118
aspersa, Helix 162
aspersum, Cornu 28, 161, **162-63**
ater, Arion 127, 128, **147**
avara, Succinea 39

Bananaslug 123, C-16-18
baskervillei, Helix 151
Bifidaria agna 53
binneyana, Nesovitrea 26, **102-3**, 103;
    Retinella 102; *Vertigo* 60, **69**
Boettgerilla pallens 27, **110-11**, C-12;
    vermiformis 110
bollisiana, Vertigo 60
breweri, Helix 98
brunnea, Polygyra townseniana 149
californica, Vertigo 57
californicum, Punctum 78
camelus, Hemphillia 28, **136-37**, C-14
campestris, Agriolimax 118; *Limax* 118
canadense, Carychium exile 38
canadica, Oreohelix strigosa 84
caputspinulae, Helix 75; *Paralaoma* 76
caruanae, Agriolimax 120
Carychium exile **38**; *magnificum* 37;
    minimum 24, **36-37**;
    occidentale 24, 36, **37-38**; *tridentatum* 37
castaneus, Helix ptychophora 148; *Limax* 118
Catinella gabbii **39**; *rehderi* **39**; *vermeta* 24, **39**
cellaria, Helix 106
cellarius, Oxychilus 27, **106-7**
Cepaea nemoralis 28, 159, **160-61**, 162, C-3
chersinella, Helix 87; *Pristiloma* 26, **87-88**
chocolata, Circinaria vanouverensis 72
Circinaria vancouverensis 72
circumscriptus, Arion 27, **129-30**, 130, 131,
    147, C-7
clappi, Planogyra 24, **44-45**, 75, 77; *Punctum* 44
Cochlicopa lubrica 24, **41-42**
cockerelli, Pyramidula 80
coeruleum, Prophysaon 28, **144-45**, C-25
columbiana, Helix 155; *Monadenia fidelis* 159;
    Polygyra 155; *Vertigo* 25, 59, **61-62**, 65;
    Vespericola 155
columbianus, Ariolimax 10, 11, 12, 18, 27,
    123-24, C-16-18; *Limax* 123;
    Vespericola 16, 28, 154, **155-56**, C-4, C-5
Columella alticola 55; *columella* 25, **54-55**, 56;
    edentula 25, 54, 55, **56**
columella, Pupa 54
columna, Retinella 88
conspecta, Helix 75
conspectum, Punctum 75, 76
contracta, Vitrea 26, **92-93**;
    Zonites crystallina 92

Conulus fulvus 94
convexior, Thysanophora ingersolli 157
cookei, Microphysula 21, 28, 93, **156-57**;
    Zonitoides 156
cooperi, Helix 85; *Oreohelix* 85
Cornu aspersum 28, 161, **162-63**
corpulenta, Isthmia 65; *Pupa* 65;
    Vertigo modesta 66
costata, Vallonia 49
crateris, Pristiloma arcticum 92
cristata, Vertigo 25, 60, **62-63**, 65;
    Vertigo gouldii 62
cronkhitei, Discus 82; *Helix* 81; *Pyramidula* 81
Cryptomastix devia 18, 28, **151-52**, 152;
    germana 18, 28, 152, **153-54**, 155;
    mullani 18, 28, 151, **152-53**
crystallina, Zonites 92
cyclophorella, Vallonia 24, **47**, 49
cylindracea, Lauria 24, **42-44**, 51; *Turbo* 42
decora, Pupa 65
Deroceras hesperium 27, **117-18**, 118, 120;
    hyperboreus 117; *laeve* 8, 9, 27, 117, **118-19**,
    120, C-21; *monentolophus* 118;
    panormitanum 27, 118, **120-21**, C-22;
    reticulatum 12, 19, 27, **121-23**, C-23
devia, Cryptomastix 18, 28, **151-52**, 152;
    Helix 151
Discus cronkhitei 82; *rotundatus* 26, **83**;
    ruderatus 82; *shimekii* 26, **80-81**, 83;
    whitneyi 18, 26, 75, 80, 81, **81-82**, 83
distinctus, Arion 27, **132-33**, 148, C-6
draparnaudi, Helix 107; *Oxychilus* 20, 27, 106,
    107-8, C-20
dromedarius, Hemphillia 28, 136, **137-38**, C-13
Earshell Slug 74-75
edentula, Columella 25, 54, 55, **56**
elatior, Vertigo 25, **63-64**
electrina, Helix 103; *Nesovitrea* 26, 102, **103-4**
elegans, Oxyloma 40
Euconulus alderi 96; *fulva* 95; *fulvus* 18, 26,
    94-95, 96; *praticola* 26, 68, 94, **95-96**
excentrica, Vallonia 24, **48-49**, 49, 51, 52
exile, Carychium **38**
eyerdami, Anguispira kochi 79
fasciatus, Arion 129, 130, 131, **147**
fidelis, Helix 159;
    Monadenia 14, 18, 28, **159-60**, 161, C-2
Fieldslugs 116-23, C-22, C-23
flavum, Prophysaon 142
flavus, Limacus 27, **112-13**, C-32; *Limax* 112
foliolatum, Prophysaon 12, 18, 28, 142, **145-47**,
    C-27
foliolatus, Arion 145
Forestsnails 148-50, C-29
frustrationensis, Allogona townsendiana 149
fulva, Euconulus 95; *Helix* 94, 95
fulvus, Conulus 94; *Euconulus* 18, 26, **94-95**, 96
gabbii, Catinella **39**
gagates, Milax **108**

Gardenslugs 112-16, C-30-32
Gardensnail 162-63
Gastrocopta holzingeri 17, 25, **53-54**, 69;
    pentodon **69**
germana, Cryptomastix 18, 28, 152, **153-54**, 155;
    Helix 20, 153; *Polygyra* 153
glandulosa, Hemphillia 28, **139-40**, C-15
Glass-snails 92-93, 101-8, 109-10, C-20, C-24
Gloss snails 98-100
gouldii, Vertigo 25, 62, 63, **64-65**
gracilicosta, Vallonia 24, 47, **49-50**
groenlandica, Succinea 39
groenlandicum, Oxyloma 24, **39-40**
Grovesnail 160-61
haliotidea, Testacella 25, **74-75**
hammonis, Nesovitrea 103
Haplotrema vancouverense 4, 25, 70, 71, **72-73**,
    C-19
harpa, Helix 45; *Zoogenetes* 10, 24, **45-46**
Hawaiia minuscula 93, **94**
hawkinsii, Oxyloma 24, **40**
hebes, Pupa 51; *Pupilla* 24, 43, **51-52**, 53
Helix alliaria 104; *arborea* 98; *aspersa* 162;
    baskervillei 151; *breweri* 98; *caputspinulae*
    75; *cellaria* 106; *chersinella* 87; *columbiana*
    155; *conspecta* 75; *cooperi* 85; *cronkhitei*
    81; *devia* 151; *draparnaudi* 107; *electrina*
    103; *fidelis* 159; *fulva* 94, 95; *germana* 20,
    153; *harpa* 45; *ingersollii* 157; *kochi* 79;
    limitaris 85; *lubrica* 41; *lucidus* 107;
    mullani 152; *nemoralis* 160; *nitida* 99;
    nitidula 101; *nuttalliana* 159; *oregonensis*
    159; *pellucida* 109; *ptychophora* 148;
    pulchella 50; *pusilla* 75; *rotundata* 83;
    servilis 75; *sportella* 70, 71; *striatella* 81;
    strigosa 84; *subrudis* 85; *townsendiana*
    149; *vancouverensis* 72; *vellicata* 72;
    whitneyi 81
hemphilli, Limax 118; *Prophysaon* 142
Hemphillia camelus 28, **136-37**, C-14;
    dromedarius 28, 136, **137-38**, C-13;
    glandulosa 28, **139-40**, C-15; *malonei* 137
hesperium, Deroceras 27, **117-18**, 118, 120
Hive snails 94-96
holzingeri, Gastrocopta 17, 25, **53-54**, 69;
    Pupa 53
hortensis, Arion 132, 133, **148**
hubrichtii, Radiodiscus 97, 98
Hyalina arctica 91; *pellucida* 103; *praticola* 95
hybrida, Selenites vancouverensis 70
hybridum, Ancotrema 25, **70-71**, 71, 72, C-33
hyperboreus, Deroceras 117; *Limax* 118
idahoensis, Pupilla muscorum 51
ingersollii, Helix 157; *Microphysula* 18, 28, 157,
    157-58; *Thysanophora* 157
intermedius, Agriolimax montanus 118;
    Arion 27, 128, **133-34**, C-11
Isthmia corpulenta 65
japonica, Pristiloma 91

johnsoni, Pristiloma 26, **88-89**; *Vitrea* 88
Jumping-slugs 136-40, C-13-15
kaibabensis, Pupilla hebes 51
kochi, Anguispira 18, 25, **79-80**, C-1; *Helix* 79
laeve, Deroceras 8, 9, 17, 117, **118-19**, 120, C-21
laevis, Limax 117, 118
lagganensis, Vertigo gouldi 63
Lancetooth snails 70-73, C-19, C-33
lansingi, Pristiloma 26, **89-90**, 90, 91; *Zonites* 89
latilabrum, Vespericola columbiana 155
Lauria cylindracea 24, **42-44**, 51
Lehmannia valentiana 27, 114, **115-16**, C-31
Limacus flavus 27, **112-13**, C-32
Limax agrestis 122; *campestris* 118;
    castaneus 118; *columbianus* 123;
    flavus 112; *hemphilli* 118; *hyperboreus* 118;
    laevis 117, 118; *marginatus* 115;
    maximus 9, 11, 12, 27, 112, **113-14**, C-30;
    montanus 118; *panormitanum* 120;
    poirieri 115; *reticulatum* 121; *rufus* 127;
    subfuscus 135; *valentiana* 115
limitaris, Helix 85
loessensis, Vertigo gouldii 63
lombardii, Allogona **156**
lubrica, Cochlicopa 24, **41-42**; *Helix* 41
lucidus, Helix 107
Macrocyclias vancouverensis 70
maculata, Ariolimax columbianus 123
magnificum, Carychium 37
Magnipelta mycophaga 18, 27, **125-26**
major, Mesodon ptycophorus 148
malonei, Hemphillia 137
marginatus, Limax 115
mariposa, Vertigo modesta 67
maximus, Limax 9, 11, 12, 27, 112, **113-14**, C-30
meridionalis, Thysanophora ingersolli 157
Mesodon ptychophorus 148; *townsendiana* 148
Microphysula cookei 21, 28, 93, **156-57**;
    ingersollii 18, 28, 157, **157-58**
Milax gagates **108**
minimum, Carychium 24, **36-37**
minor, Mesodon townsendiana 148
minuscula, Hawaiia 93, **94**
minutissimum, Punctum 78
modesta, Vertigo 25, 60, 62, **65-66**, 67
Monadenia fidelis 14, 18, 28, **159-60**, 161, C-2
monentolophus, Deroceras 118
montana, Vallonia costata 49
montanus, Agriolimax 118; *Limax* 118
mullani, Cryptomastix 18, 28, 151, **152-53**;
    Helix 152; *Polygyra* 152
muscorum, Pupilla 51, **53**
mycophaga, Magnipelta 18, 27, **125-26**
Nearctula rowellii 57; *species* 25, **57**, 59
nefas, Pupilla hebes 51
nemoralis, Cepaea 28, 159, **160-61**, 162, C-3;
    Helix 160
Nesovitrea binneyana 26, **102-3**, 103;
    electrina 26, 102, **103-4**; *hammonis* 103

niger, Ariolimax columbianus 123
nitida, Helix 99
nitidula, Aegopinella 20, 26, 98, 99, **101-2**, 104;
    Helix 101
nitidus, Zonitoides 8, 26, 68, 98, **99-100**, 105
nuttalliana, Helix 159
nuttallianum, Oxyloma 24, **40**
obscurum, Prophysaon 141, 142
occidentale, Carychium 24, 36, **37-38**
occidentalis, Anguispira kochi 80; *Limax*
    campestris 118; *Nesovitrea binneyana* 103;
    Patula solitaria 79; *Retinella binneyana* 102
olneyae, Cryptomastix mullani 153;
    Polygyra mullani 152
Oregonian snails 151-54
oregonensis, Helix 159; *Succinea* 40
Oreohelix cooperi 85; *strigosa* 26, **84-85**, 86;
    subrudis 26, 85, **85-86**
ovata, Pupa 67; *Vertigo* 25, 63, **67-68**, 68, 69
owenii, Arion 133
Oxychilus alliarius 11, 27, **104-5**, C-24, 106;
    cellarius 27, **106-7**;
    draparnaudi 20, 27, 106, **107-8**, C-20
Oxyloma elegans 40; *groenlandicum* 24, **39-40**;
    hawkinsii 24, **40**; *nuttallianum* 24, **40**;
    pfeifferi 40
pacificum, Prophysaon 142
pallens, Boettgerilla 27, **110-11**, C-12
panormitanum, Deroceras 27, 118, **120-21**, C-22;
    Limax 120
Paralaoma caputspinulae 76; *servilis* 25, 44, 47,
    75-76, 77, 82
parietalis, Pupa corpulenta 65;
    Vertigo modesta 66
pasadenae, Punctum conspectum 75
Patula solitaria 79
Patulastra pugetensis 97
pellucida, Helix 109; *Hyalina* 103;
    Vitrina 7, 27, **109-10**
pentodon, Gastrocopta **69**
pfeifferi, Oxyloma 40; *Vitrina* 109
pictus, Limax hemphilli 118
pilosa, Polygyra columbiana 155
Planogyra clappi 24, **44-45**, 75, 77
poirieri, Limax 115
Polygyra columbiana 155; *germana* 153;
    mullani 152; *townsendiana* 149
praticola, Euconulus 26, 68, 94, **95-96**;
    Hyalina 95
Pristiloma arcticum 26, 90, **91-92**;
    chersinella 26, **87-88**; *japonica* 91;
    johnsoni 26, **88-89**; *lansingi* 26, **89-90**, 90,
    91; *stearnsii* 26, **90-91**; *taylori* 88
Prophysaon andersoni 9, 19, 28, **142-43**, 145,
    C-26; *coeruleum* 28, **144-45**, C-25;
    flavum 142; *foliolatum* 12, 18, 28, 142, **145-
    47**, C-27; *hemphilli* 142; *obscurum* 141,
    142; *pacificum* 142; *vanattae* 28, **141-42**,
    143, C-28

ptychophora, Allogona 18, 28, 79, **148-49**, 149;
  Helix 148
ptychophorus, Mesodon 148
pugetensis, Patulastra 97; Striatura 18, 26, 44,
  97-98
pulchella, Helix 50; Vallonia 24, 48, 49, **50-51**
Punctum californicum 78; clappi 44;
  conspectum 75, 76; minutissimum 78;
  pygmaeum 77, 78; randolphii 25, 43, 44, 56,
  75, **77-78**
Pupa columella 54; corpulenta 65; decora 65;
  hebes 51; holzingeri 53; ovata 67
Pupilla alticola 54; hebes 24, 43, **51-52**, 53;
  muscorum 51, **53**
pusilla, Helix 75
pygmaeum, Punctum 77, 78
Pyramidula cockerelli 80; cronkhitei 81;
  randolphii 77
Radiodiscus hubrichtii 97, 98
randolphi, Zonitoides 80, 81
randolphii, Punctum 25, 43, 44, 56, 75, **77-78**;
  Pyramidula 77
rehderi, Catinella **39**
reticulatum, Deroceras 12, 19, 27, **121-23**, C-23;
  Limax 121
Retinella binneyana 102; columna 88
rotundata, Helix 83
rotundatus, Discus 26, **83**
rowellii, Nearctula 57; Vertigo 57
ruderatus, Discus 82
rufus, Arion 27, **127-28**, 133, 147, C-8, C-9;
  Limax 127
rusticana, Succinea 40
sanbernardinensis, Vertigo andrusiana 60
Selenites vacouverensis 70
semialba, Monadenia fidelis 159
semidecussata, Macrocyclis vancouverensis 70
septuagentaria, Vallonia cyclophorella 47
servilis, Helix 75;
  Paralaoma 25, 44, 47, **75-76**, 77, 82
shimekii, Discus 26, **80-81**, 83; Zonites 80
silvaticus, Arion 27, 130, **130-31**
sinkyonum, Ancotrema sportella 72
solitaria, Patula 79
sonorana, Vallonia 49
sportella, Ancotrema 25, 71, **71-72**; Helix 70, 71
stantoni, Oreohelix strigosa 84
stearnsii, Pristiloma 26, **90-91**; Zonites 90
steindachneri, Ariolimax 123
stramineus, Ariolimax columbianus 124
striatella, Helix 81
Striatura pugetensis 18, 26, 44, **97-98**
strigosa, Helix 84; Oreohelix 26, **84-85**, 86
subfuscus, Arion 4, 7, 28, **135-36**, C-10;
  Limax 135
subrudis, Helix 85; Oreohelix 26, 85, **85-86**
Succinea avara 39; groenlandica 39;
  oregonensis 40; rusticana 40; verrilli 39

Taildropper slugs 141-46, C-25-28
taylori, Pristiloma 88
Testacella haliotidea 25, **74-75**
Thorn snails 36-38
Thysanophora igersolli 157
Tigersnail 79-80, C-1
Tightcoil snails 87-92
townsendiana, Allogona 18, 28, 148, **149-50**,
  C-29; Helix 149; Mesodon 148;
  Polygyra 149
tridentatum, Carychium 37
tristis, Agriolimax montanus 118
Turbo cylindracea 42
typica, Ariolimax columbianus 123
typicus, Agriolimax montanus 118
upsoni, Zonites 67
valentiana, Lehmannia 27, 114, **115-16**, C-31;
  Limax 115
Vallonia albula 49, 50; costata 49
cyclophorella 24, **47**, 49; excentrica 24, **48-49**,
  49, 51, 52; gracilicosta 24, 47, **49-50**;
  pulchella 24, 48, 49, **50-51**; sonorana 49
vanattae, Prophysaon 28, **141-42**, C-28
vancouverense, Haplotrema 4, 25, 70, 71, **72-73**,
  C-19
vancouverensis, Circinaria 72; Helix 72;
  Macrocyclias 70; Selenites 70
vancouverinsulae, Polygyra germana 153
vellicata, Helix 72
ventricosa, Vertigo 63, 69
vermeta, Catinella 24, **39**
vermiformis, Boettgerilla 110
verrilli, Succinea 39
Vertigo andrusiana 25, **59-60**, 61; arctica 66;
  arthuri 25, **60-61**, 64, 65; binneyana 60, **69**;
  bollisiana 60; californica 57; columbiana
  25, 59, **61-62**, 65; cristata 25, 60, **62-63**, 65;
  elatior 25, **63-64**; gouldi/gouldii 25, 62, 63,
  64-65; modesta 25, 60, 62, **65-66**, 67; ovata
  25, 63, **67-68**, 68, 69; rowellii 57; species
  25, 63, **68-69**; ventricosa 63, 69
Vespericola columbianus 16, 28, 154, **155-56**,
  C-4, C-5
Vitrea contracta 26, **92-93**; johnsoni 88
Vitrina alaskana 109, 110; angelicae 109;
  pellucida 7, 27, **109-10**; pfeifferi 109
whitneyi, Discus 18, 26, 75, 80, 81, **81-82**, 83;
  Helix 81
Wormslug 110-11, C-12
zonatipes, Agriolimax campestris 118
Zonites crystallina 92; lansingi 89; shimekii
  80; stearnsii 90; upsoni 67
Zonitoides arboreus 18, 26, **98-99**, 99, 101, 105;
  cookei 156; nitidus 8, 26, 68, 98, **99-100**,
  105; randolphi 80, 81
Zoogenetes harpa 10, 24, **45-46**